Non-Equilibrium Particle Dynamics

Edited by Albert S. Kim

Published in London, United Kingdom

IntechOpen

Supporting open minds since 2005

Non-Equilibrium Particle Dynamics
http://dx.doi.org/10.5772/intechopen.78729
Edited by Albert S. Kim

Contributors
Serhiy Vladimirovich Bobyr, Gerd Roepke, Sten Sarman, Aatto Laaksonen, Yonglei Wang, Aamir Shahzad, Maogang He, Jorge Fernando Camacho, Rosalio Rodriguez, William Uspal, Albert S. Kim

First published in London, United Kingdom, 2019 by IntechOpen
IntechOpen is the global imprint of INTECHOPEN LIMITED, registered in England and Wales, registration number: 11086078, 7th floor, 10 Lower Thames Street, London, EC3R 6AF, United Kingdom
Printed in Croatia

British Library Cataloguing-in-Publication Data
A catalogue record for this book is available from the British Library

Additional hard and PDF copies can be obtained from orders@intechopen.com

Non-Equilibrium Particle Dynamics
Edited by Albert S. Kim
p. cm.
Print ISBN 978-1-83968-077-9
Online ISBN 978-1-83968-078-6
eBook (PDF) ISBN 978-1-83968-079-3

For EU product safety concerns:
IN TECH d.o.o., Prolaz Marije Krucifikse Kozulić 3, 51000 Rijeka, Croatia, info@intechopen.com or visit our website at intechopen.com.

Meet the editor

Dr. Albert S. Kim is currently a full professor at the Department of Civil and Environmental Engineering at the University of Hawaii at Manoa. He has been teaching in the department from 2001, after he earned his MS (1997) and PhD (2000) in Civil and Environmental Engineering from the University of California at Los Angeles. Dr. Kim's scientific accomplishments include the National Science Foundation Faculty Early Career (CAREER) Award in 2005, the University of Hawaii Regents' Medal for Excellence in Research in 2006, and the University of Hawaii Regents' Medal for Excellence in Teaching in 2017. Professor Kim has published more than 60 peer-reviewed journal papers and six book chapters, and edited two open access books.

Contents

Preface

Thermodynamics is used to describe the engineering processes of various disciplines such as chemical, environmental, and mechanical engineering. Fundamentally, thermodynamics deals with heat and temperature to mechanical work and energy from a macroscopic viewpoint. Statistical mechanics explains macroscopic behaviors of a thermodynamic system using entropy S, energy E, matter number N, temperature T, chemical potential μ, volume V, and pressure P, and the thermodynamic second law relates the internal E, V, and N. In statistical mechanics, an ensemble denotes a set of three independent variables to be used to characterize a system of interest, and ensembles are classified into isothermal (for constant temperature) and adiabatic (equivalent to isentropic, i.e., of constant entropy) ones. Popular isothermal ensembles include canonical (NVT), grand canonical (μVT), and isothermal/isobaric (NPT) ensembles. A representative adiabatic ensemble is a microcanonical ensemble using N, V, and E as primary variables, which is less popular for statistical simulations due to the inconvenient constraint of fixed energy value. Theoretically, eight ensembles (four isothermal and four adiabatic) can be made, but in principle only seven ensembles are available, because the last isothermal ensembles use only intensive variables of μ, P, and T, which are not enough to decide a system scale. More importantly, an ensemble presumes an equilibrium state of an isolated system, which does not allow mass and heat exchange to the surroundings.

Other sub-branches of theoretical physics are linked to statistical mechanics through thermodynamic variables: classical mechanics through temperature T, fluid mechanics through pressure P, and chemistry through chemical potential μ. Although equilibrated systems can be fully analyzed using a thermodynamic ensemble, most of the engineering processes consist of open systems through which mass, heat, and momentum can be exchanged from the system interior to surroundings. In principle an open system can reach a steady state, which is irreversible or of non-equilibrium. The steady state can be mathematically represented as $\partial/\partial t$ // 0, which means no physical quantities associated with the system explicitly change with time. In equilibrium, the entropy is already maximized and therefore it remains constant. In the steady state, changing rates of physical variables can be constant so that entropy steadily increases with respect to time. If the system of interest is open, the energy usually dissipates and hence entropy increases. Energy dissipation occurs due not only to the openness of the system but also to an incomplete conversion of a type of energy to the other type. The inelastic nature of materials within the system converts kinetic energy to thermal energy, often lost to the surrounding environment. To accurately analyze an engineering process, understanding underlying phenomena using irreversible thermal balance equations is essential. In this vein, this book covers various aspects of irreversible statistical mechanics and non-equilibrium thermodynamics as a partial contribution to the irreversible statistical mechanics in theoretical physics.

Albert S. Kim
University of Hawaii at Manoa,
USA

Section 1

Irreversible Statistical Mechanics

Chapter 1

Nonequilibrium Statistical Operator

Gerd Röpke

Abstract

Nonequilibrium statistical physics is concerned with a fundamental problem in physics, the phenomenon of irreversibility, which is not rigorously solved yet. Different approaches to the statistical mechanics of nonequilibrium processes are based on empirical assumptions, but a rigorous, first principle theory is missing. An important contribution to describe irreversible behavior starting from reversible Hamiltonian dynamics was given by Zubarev, who invented the method of the nonequilibrium statistical operator (NSO). We discuss, in particular, the extended von Neumann equation and the entropy concept in this approach. The method of NSO proved to be a general and universal approach to different nonequilibrium phenomena. Typical applications are the quantum master equation, kinetic theory, and linear response theory which are outlined and illustrated solving standard examples for reaction and transport processes. Some open questions are emphasized.

Keywords: Zubarev method, NSO, quantum master equation, kinetic equations, linear response theory

1. Introduction: irreversible processes

1.1 Irreversibility and arrow of time

Irreversibility belongs to the unsolved fundamental problems in recent physics. Nonequilibrium processes are omnipresent in our daily experience. However, a fundamental, microscopic description of such processes is missing yet.

Our microscopic description of physical phenomena is expressed by equations of motion, well known in mechanics, electrodynamics, quantum mechanics, and field theory. We model a physical system, we determine the degrees of freedom and the forces, and we introduce a Lagrangian. The equations of motion are differential equations. If we know the initial state, the future of the system can be predicted solving the equations of motion. Anything is determined. The equations of motion are invariant with respect to time reversion. The time evolution is reversible. No arrow of time is selected out, nothing happens what is not prescribed by the initial state.

This picture was created by celestial dynamics. It is very successful, very presumptuous, and many processes are described with high precision. However, it is in contradiction to daily experience. We know birth and death, decay, destruction, and many other phenomena that are irreversible, selecting out the arrow of time.

A qualitative new discipline in physics is thermodynamics. It considers not a model but any real system. The laws of thermodynamics define new quantities, the state variables. The second law determines the entropy S as state variable (and the temperature T) via

$$dS = \frac{1}{T}\delta Q_{\text{reversible}} \tag{1}$$

where δQ is the heat imposed to the system within a reversible process, together with the third law which fixes the value $S(T = 0) = 0$ independent on other state variables. For irreversible processes holds

$$\frac{dS}{dt} > \frac{\delta Q}{T}. \tag{2}$$

In particular, for isolated system, $\delta Q = 0$, irreversible processes are possible so that $dS/dt>0$. Typical examples are friction that transforms mechanical energy to thermal energy, temperature equilibration without production of mechanical work, diffusion processes to balance concentration gradients. An arrow of time is selected out, time reversion describes a phenomenon which is not possible. How can irreversible evolution in time be obtained from the fundamental microscopic equations of motion which are reversible in time?

For equilibrium thermodynamics, a microscopic approach is given by statistical physics. Additional concepts are introduced such as probability and distribution function, ensembles in thermodynamic equilibrium, and information theory. New thermodynamic quantities are introduced, basically the entropy, which have no direct relation to mechanical quantities describing the equation of motion. However, nonequilibrium processes are described in a phenomenological way, and no fundamental solution of the problem of irreversibility is found until now. A substantial step to develop a theory of irreversible evolution is the Zubarev method of the nonequilibrium statistical operator (NSO) [1–6] to be described in the following section. It is a consistent theory to describe different nonequilibrium processes what is indispensable for a basic approach.

1.2 Langevin equation

To give an example for a microscopic approach to a nonequilibrium process, let us consider the Brownian motion. A particle suspended in a liquid, moving with velocity $\mathbf{v}_{\text{medium}}$, experiences a friction force $\mathbf{F}^{\text{fric}}(t)$,

$$\frac{d}{dt}\mathbf{v}(t) = \frac{1}{m}\mathbf{F}^{\text{fric}}(t) = -\gamma[\mathbf{v}(t) - \mathbf{v}_{\text{medium}}], \tag{3}$$

with the coefficient of friction γ. The solution

$$\mathbf{v}(t) = \mathbf{v}(t_0)e^{-\gamma(t-t_0)} + \mathbf{v}_{\text{medium}}\left[1 - e^{-\gamma(t-t_0)}\right] \tag{4}$$

describes the relaxation from the initial state $\mathbf{v}(t_0)$ at t_0 to the final state $\mathbf{v}_{\text{medium}}$ for $t - t_0 \to \infty$. Independent of the initial state, the particle rests in equilibrium with the medium. In the general case not considered here, an external force can be added.

As it is well known, this simple relaxation behavior cannot be correct because it does not describe the Brownian motion, showing the existence of fluctuations also in thermal equilibrium. This problem was solved with the Langevin equation: instead of the trajectory $\mathbf{v}(t)$ as solution of a differential equation, the stochastic process $\mathbf{V}(t)$ is considered. It obeys the stochastic differential equation

$$\frac{d}{dt}\mathbf{V}(t) = -\gamma[\mathbf{V}(t) - \mathbf{v}_{\rm rel}(t)] + \mathbf{R}(t). \tag{5}$$

The random acceleration $\mathbf{R}(t)$ (or the stochastic force $m\mathbf{R}(t)$) is a stochastic process, which is characterized by special properties. For instance, white noise is a Gaussian process that is characterized by the mean value $\langle \mathbf{R}(t) \rangle = 0$ and the autocorrelation function

$$\langle R_i(t_1) R_j(t_2) \rangle = \varphi_{ij}(t_2 - t_1) = 2D\delta_{ij}\delta(t_2 - t_1). \tag{6}$$

D is the diffusion coefficient. An interesting result is the Einstein relation (fluctuation-dissipation theorem, FDT)

$$\frac{D}{\gamma} = \frac{k_{\rm B}T}{m} \tag{7}$$

which relates the friction coefficient γ (dissipation) to the fluctuations φ in the system (stochastic forces), characterized by the parameter D; see [5] for more details.

1.3 Von Neumann equation

Within statistical mechanics, the thermodynamic state of an ensemble of many-particle systems at time t is described by the statistical operator $\rho(t)$. We assume that the time evolution of the quantum state of the system is given by the Hamiltonian H^t which may contain time-dependent external fields. The von Neumann equation follows as equation of motion for the statistical operator,

$$\frac{\partial}{\partial t}\rho(t) + \frac{\mathrm{i}}{\hbar}[\mathrm{H}^t, \rho(t)] = 0. \tag{8}$$

The von Neumann equation describes reversible dynamics. The equation of motion is based on the Schrödinger equation. Time inversion and conjugate complex means that the first term on the left-hand side as well as the second one change the sign, since $\mathrm{i} \to -\mathrm{i}$ and both the Hamiltonian and the statistical operator are Hermitean. However, the von Neumann equation is not sufficient to determine $\rho(t)$ because it is a first-order differential equation, and an initial value $\rho(t_0)$ at time t_0 is necessary to specify a solution. This problem emerges clearly in equilibrium.

1.4 Thermodynamic equilibrium and entropy

By definition, in thermodynamic equilibrium, the thermodynamic state of the system is not changing with time. Both, H^t and $\rho(t)$, are not depending on t so that

$$\frac{\partial}{\partial t}\rho_{\rm eq}(t) = 0. \tag{9}$$

The solution of the von Neumann equation in thermodynamic equilibrium becomes trivial, $\frac{\mathrm{i}}{\hbar}\left[\mathrm{H}, \rho_{\rm eq}\right] = 0$. The time-independent statistical operator $\rho_{\rm eq}$

commutes with the Hamiltonian. We conclude that ρ_{eq} depends only on constants of motion C_n that commute with H. But, the von Neumann equation is not sufficient to determine how ρ_{eq} depends on constants of motion C_n. We need a new additional principle, not included in Hamiltonian dynamics.

Equilibrium statistical mechanics is based on the following principle to determine the statistical operator ρ_{eq}: consider the functional (information entropy)

$$S_{\mathrm{inf}}[\rho] = -\mathrm{Tr}\{\rho \ln \rho\} \tag{10}$$

for arbitrary ρ that is consistent with the given conditions $\mathrm{Tr}\{\rho\} = 1$ (normalization) and

$$\mathrm{Tr}\{\rho \mathrm{C}_n\} = \langle \mathrm{C}_n \rangle \tag{11}$$

(self-consistency conditions). Respecting these conditions, we vary ρ and determine the maximum of the information entropy for the optimal distribution ρ_{eq} so that $\delta S_{\mathrm{inf}}[\rho_{\mathrm{eq}}] = 0$. As it is well-known, the method of Lagrange multipliers can be used to account for the self-consistency conditions (11). The corresponding maximum value for $S_{\mathrm{inf}}[\rho]$

$$S_{\mathrm{eq}}\left[\rho_{\mathrm{eq}}\right] = -k_{\mathrm{B}}\mathrm{Tr}\left\{\rho_{\mathrm{eq}} \ln \rho_{\mathrm{eq}}\right\} \tag{12}$$

is the equilibrium entropy of the system at given constraints $\langle \mathrm{C}_n \rangle$ and k_{B} is the Boltzmann constant. The solution of this variational principle leads to the Gibbs ensembles for thermodynamic equilibrium, see also Section 4.

As an example, we consider an open system which is in thermal contact and particle exchange with reservoirs. The corresponding equilibrium statistical operator has to obey the given constraints: normalization $\mathrm{Tr}\{\rho\} = 1$, thermal contact with the bath so that $\mathrm{Tr}\{\rho \mathrm{H}\} = U$ (internal energy), particle exchange with a reservoir so that for the particle number operator N_c of species c, the average is given by $\mathrm{Tr}\{\rho \mathrm{N}_c\} = n_c \Omega$, where Ω denotes the volume of the system (we do not use V to avoid confusion with the potential), and n_c is the particle density of species c. Looking for the maximum of the information entropy functional with these constraints, one obtains the grand canonical distribution

$$\rho_{\mathrm{eq}} = \frac{e^{-\beta\left(\mathrm{H} - \sum_c \mu_c \mathrm{N}_c\right)}}{\mathrm{Tr}e^{-\beta\left(\mathrm{H} - \sum_c \mu_c \mathrm{N}_c\right)}}. \tag{13}$$

The normalization is explicitly accounted for by the denominator (partition function). The second condition means that the energy of a system in heat contact with a thermostat fluctuates around an averaged value $\langle \mathrm{H} \rangle = U = u\Omega$ with the given density of internal energy u. This condition is taken into account by the Lagrange multiplier β that must be related to the temperature, a more detailed discussion leads to $\beta = 1/(k_{\mathrm{B}}T)$. Similarly, the contact with the particle reservoir fixes the particle density n_c, introduced by the Lagrange multiplier μ_c, which has the meaning of the chemical potential of species c.

Within the variational approach, the Lagrange parameters β, μ_c have to be eliminated. This leads to the equations of state ($\langle ... \rangle_{\mathrm{eq}} = \mathrm{Tr}\{\rho_{\mathrm{eq}} ...\}$) which relate, e.g., the chemical potentials μ_c to the particle densities n_c,

$$\langle \mathrm{H} \rangle_{\mathrm{eq}} = U(\Omega, \beta, \mu_c), \qquad \langle \mathrm{N}_c \rangle_{\mathrm{eq}} = \Omega n_c(T, \mu_c). \tag{14}$$

The entropy $S_{eq}(\Omega, \beta, \mu)$ follows from Eq. (12). The dependence of extensive quantities on the volume Ω is trivial for homogeneous systems. After a thermodynamic potential is calculated, all thermodynamic variables are derived in a consistent manner. The method to construct statistical ensembles from the maximum of entropy at given conditions, which take into account the different contacts with the surrounding bath, is well accepted in equilibrium statistical mechanics and is applied successfully to different phenomena, including phase transitions.

Can we extend the definition of equilibrium entropy (12) also for $\rho(t)$ that describes the evolution in nonequilibrium? Time evolution is given by an unitary transformation that leaves the trace invariant. Thus, the expression $\mathrm{Tr}\{\rho(t)\ln\rho(t)\}$ is constant for a solution $\rho(t)$ of the von Neumann equation

$$\frac{\mathrm{d}}{\mathrm{d}t}[\mathrm{Tr}\{\rho(t)\ln\rho(t)\}] = 0. \tag{15}$$

The entropy for a system in nonequilibrium, however, may increase with time, according to the second law of thermodynamics. The equations of motion, including the Schrödinger equation and the Liouville-von Neumann equation, describe reversible motion and are not appropriate to describe irreversible processes. Therefore, the entropy concept (12) elaborated in equilibrium statistical physics together with the Liouville-von Neumann equation cannot be used as fundamental approach to nonequilibrium statistical physics.

2. The method of nonequilibrium statistical operator (NSO)

After the laws of thermodynamics have been formulated in the nineteenth century, in particular, the definition of entropy for systems in thermodynamic equilibrium and the increase of intrinsic entropy in nonequilibrium processes, a microscopic approach to nonequilibrium evolution was first given by Ludwig Boltzmann who formulated the kinetic theory of gases [7] using the famous Stoßzahlansatz. The question how irreversible evolution in time can be obtained from reversible microscopic equations has been arisen immediately and was discussed controversially.

The rigorous derivation of the kinetic equations from a microscopic description of a system was given only a long time afterward by Bogoliubov [8] introducing a new additional theorem, the principle of weakening of initial correlation.

2.1 Construction of the Zubarev NSO

A generalization of this principle has been given by Zubarev [1, 2], who invented the method of the nonequilibrium statistical operator (NSO). This approach has been applied to various problems in nonequilibrium statistical physics, see [3, 4] and may be considered as a unified, fundamental approach to nonequilibrium systems which includes different theories such as quantum master equations, kinetic theory, and linear response theory to be outlined below. An exhaustive review of the Zubarev NSO method and its manifold applications cannot be given here, see [1–5].

In the first step, we interrogate the concept of thermodynamic equilibrium. This is an idealization, because slow processes are always possible. As example, we may take the nuclear decay of long-living isotopes, hindered chemical reactions, or the long-time relaxation of glasses. Concepts introduced for equilibrium have to be generalized to nonequilibrium. An example is the thermodynamics of irreversible processes.

2.1.1 The relevant statistical operator

A solution of the problem to combine equilibrium thermodynamics and nonequilibrium processes was proposed by Zubarev [1, 2]. To characterize the nonequilibrium state of a system, we introduce the set of relevant observables $\{B_n\}$ extending the set of conserved quantities $\{C_n\}$. At time t, the observed values $\langle B_n \rangle^t$ have to be reproduced by the statistical operator $\rho(t)$, i.e.,

$$\mathrm{Tr}\{\rho(t)\, B_n\} = \langle B_n \rangle^t. \tag{16}$$

However, these conditions are not sufficient to fix $\rho(t)$, and we need an additional principle to find the correct one in between many possible distributions which all fulfill the conditions (16). We ask for the most probable distribution at time t, where the information entropy has a maximum value (see Section 4)

$$\delta[\mathrm{Tr}\{\rho_{\mathrm{rel}}(t) \ln \rho_{\mathrm{rel}}(t)\}] = 0 \tag{17}$$

with the self-consistency conditions

$$\mathrm{Tr}\{\rho_{\mathrm{rel}}(t) B_n\} = \langle B_n \rangle^t \tag{18}$$

and $\mathrm{Tr}\{\rho_{\mathrm{rel}}(t)\} = 1$. Once more, we use Lagrange multipliers $\lambda_n(t)$ to account for the self-consistency conditions (18). Since the averages are, in general, time dependent, the corresponding Lagrange multipliers are now time-dependent functions as well. We find the generalized Gibbs distribution

$$\rho_{\mathrm{rel}}(t) = \mathrm{e}^{-\Phi(t) - \sum_n \lambda_n(t) B_n}, \qquad \Phi(t) = \ln\, \mathrm{Tr}\left\{ \mathrm{e}^{-\sum_n \lambda_n(t) B_n} \right\}, \tag{19}$$

where the Lagrange multipliers $\lambda_n(t)$ (thermodynamic parameters) are determined by the self-consistency conditions (18). $\Phi(t)$ is the Massieux-Planck function, needed for normalization purposes and playing the role of a thermodynamic potential. Generalizing the equilibrium case, Eq. (12), we can consider the *relevant entropy in nonequilibrium*

$$S_{\mathrm{rel}}(t) = -k_{\mathrm{B}}\, \mathrm{Tr}\, \{\rho_{\mathrm{rel}}(t) \ln \rho_{\mathrm{rel}}(t)\}. \tag{20}$$

Relations similar to the relations known from equilibrium thermodynamics can be derived. In particular, the production of entropy results as

$$\frac{\partial S_{\mathrm{rel}}(t)}{\partial t} = \sum_n \lambda_n(t) \langle \dot{B}_n \rangle^t \tag{21}$$

as known from the thermodynamics of irreversible processes. In contrast to Eq. (15), this expression can have a positive value so that $S_{\mathrm{rel}}(t)$ can increase with time.

The relevant statistical operator $\rho_{\mathrm{rel}}(t)$ is not the wanted nonequilibrium statistical operator $\rho(t)$ because it does not obey the Liouville-von Neumann equation. Also, $S_{\mathrm{rel}}(t)$ is not the thermodynamic entropy because it is based on the arbitrary choice of the set $\{B_n\}$ of relevant observables, and not all possible variables are correctly reproduced. As example, we consider below the famous Boltzmann entropy that is based on the single-particle distribution function, but does not take higher order correlation functions into account.

2.2 The Zubarev solution of the initial value problem

The solution of the problem how to find the missing signatures of $\rho(t)$ not already described by $\rho_{\rm rel}(t)$ was found by Zubarev [1, 2] generalizing the Bogoliubov principle of weakening of initial correlations [8]. He proposed to use the relevant statistical operator $\rho_{\rm rel}(t_0)$ at some initial time t_0 as initial condition to construct $\rho(t)$,

$$\rho_{t_0}(t) = \mathrm{U}(t,t_0)\rho_{\rm rel}(t_0)\mathrm{U}^\dagger(t,t_0). \tag{22}$$

The unitary time evolution operator $\mathrm{U}(t,t_0)$ is the solution of the differential equation

$$i\hbar\frac{\partial}{\partial t}\mathrm{U}(t,t_0) = \mathrm{H}^t\mathrm{U}(t,t_0), \tag{23}$$

with the initial condition $\mathrm{U}(t_0,t_0) = 1$. This unitary operator is known from the solution of the Schrödinger equation. If the Hamiltonian is not time dependent, we have

$$\mathrm{U}(t,t_0) = \mathrm{e}^{-\frac{i}{\hbar}\mathrm{H}(t-t_0)}. \tag{24}$$

If the Hamiltonian is time dependent, the solution is given by a time-ordered exponent.

Now, it is easily shown that $\rho_{t_0}(t)$ is a solution of the von Neumann equation. All missing correlations not contained in $\rho_{\rm rel}(t_0)$ are formed dynamically during the time evolution of the system. However, incorrect initial correlations contained in $\rho_{\rm rel}(t_0)$ may survive for a finite time interval $t - t_0$, and the self-consistency conditions (18) valid at t_0 are not automatically valid also at t.

To get rid of these incorrect initial correlations, according to the Bogoliubov principle of weakening of initial correlations, one can consider the limit $t_0 \to -\infty$. According to Zubarev, it is more efficient to average over the initial time so that no special time instant t_0 is singled out. This is of importance, for instance, if there are long-living oscillations determined by the initial state. According to Abel's theorem, see [1–4], the limit $t_0 \to -\infty$ can be replaced by the limit $\epsilon \to +0$ in the expression

$$\rho_\epsilon(t) = \epsilon\int_{-\infty}^{t}\mathrm{e}^{\epsilon(t_1-t)}\mathrm{U}(t,t_1)\rho_{\rm rel}(t_1)\mathrm{U}^\dagger(t,t_1)\mathrm{d}t_1. \tag{25}$$

This averaging over different initial time instants means a mixing of phases so that long-living oscillations are damped out. Finally, we obtain the nonequilibrium statistical operator as

$$\rho_{\rm NSO}(t) = \lim_{\epsilon\to 0}\rho_\epsilon(t). \tag{26}$$

This way, $\rho_{\rm rel}(t_1)$ for all times $-\infty < t_1 < t$ serves as initial condition to solve the Liouville-von Neumann equation, according to the Bogoliubov principle of weakening of initial correlations. The missing correlations are formed dynamically during the time evolution of the system. The more information about the nonequilibrium state are used to construct the relevant statistical operator, the less dynamical formation of the correct correlations in $\rho(t)$ is needed. The limit $t_0 \to -\infty$ is less active to produce the remaining missing correlating. The past that is of relevance, given by the relaxation time τ, becomes shorter, if the relevant (long-living) correlations are already correctly implemented. The limit $\varepsilon \to +0$ has to be performed after the thermodynamic limit, see below.

2.3 Discussion of the Zubarev NSO approach

2.3.1 The extended Liouville-von Neumann equation

The nonequilibrium statistical operator $\rho_\epsilon(t)$, Eq. (25), obeys the extended von Neumann equation

$$\frac{\partial \rho_\epsilon(t)}{\partial t} + \frac{\mathrm{i}}{\hbar}[\mathrm{H}^t, \rho_\epsilon(t)] = -\epsilon(\rho_\epsilon(t) - \rho_{\mathrm{rel}}(t)). \tag{27}$$

as can be seen after simple derivation with respect to time. In contrast to the von Neumann equation (8), a source term arises on the right-hand side that becomes infinitesimal small in the limit $\epsilon \to +0$. This source term breaks the time inversion symmetry, so that for any finite value of ϵ, the solution $\rho_\epsilon(t)$ describes in general an irreversible evolution with time.

The source term can be interpreted in the following way:

1. The source term implements the "initial condition" in the equation of motion as expressed by $\rho_{\mathrm{rel}}(t)$. Formally, the source term looks like a relaxation process. In addition to the internal dynamics, the system evolves toward the relevant distribution.

2. The construction of the source term is such that the time evolution of the relevant variables is not affected by the source term (we use the invariance of the trace with respect to cyclic permutations),

$$\frac{\partial}{\partial t}\langle \mathrm{B}_n\rangle^t = \mathrm{Tr}\left\{\frac{\partial \rho_\epsilon(t)}{\partial t}\mathrm{B}_n\right\} = -\mathrm{Tr}\left\{\frac{\mathrm{i}}{\hbar}[\mathrm{H}^t, \rho_\epsilon(t)]\mathrm{B}_n\right\} = \left\langle \frac{\mathrm{i}}{\hbar}[\mathrm{H}^t, \mathrm{B}_n]\right\rangle^t = \langle \dot{\mathrm{B}}_n\rangle^t. \tag{28}$$

The source term cancels because of the self-consistency conditions (18). Thus, the time evolution of the relevant observables satisfies the dynamical equations of motion according to the Hamiltonian H^t.

3. The value of ϵ has to be small enough, $\epsilon \ll 1/\tau$, so that all relaxation processes to establish the correct correlations, i.e., the correct distribution of the irrelevant observables, can be performed. However, $\hbar\epsilon$ has to be large compared to the energy difference of neighbored energy eigenstates of the system so that mixing is possible. For a system of many particles, the density of energy eigenvalues is high so that we can assume a quasi-continuum. This is necessary to allow for dissipation. The van Hove limit means that the limit $\epsilon \to +0$ has to be performed after the thermodynamic limit.

4. Differential equations can have degenerated solutions. For instance, we know the retarded and advanced solution of the wave equation that describes the emission of electromagnetic radiation. An infinitesimal small perturbation can destroy this degeneracy and select out a special solution, here the retarded one.

Similar problems are known for systems (magnetism) where the ground state has a lower symmetry than the Hamiltonian.

5. Any real system is in contact with the surroundings. The intrinsic dynamics described by the Hamiltonian H^t is modified due to the coupling of the open system to the bath. Within the quantum master equation approach, we can approximate the influence term describing the coupling to the bath by a relaxation term such as the source term. At present, we consider the source term as a purely mathematical tool to select the retarded solution of the von Neumann equation, and physical results are obtained only after performing the limit $\epsilon \to 0$.

2.3.2 Selection of the set of relevant observables

The Zubarev method to solve the initial value problem for the Liouville-von Neumann equation is based on the selection of the set $\{B_n\}$ of relevant observables which characterize the nonequilibrium state. The corresponding relevant statistical operator $\rho_{\rm rel}(t)$ is some approximation to $\rho(t)$. According to the Bogoliubov principle of weakening of initial correlations, the missing correlations to get $\rho(t)$ are produced dynamically. This process, the dynamical formation of the missing correlations, needs some relaxation time τ. If we would take instead of $\rho_{\rm rel}(t)$ the exact (but unknown) solution $\rho(t)$, the relaxation time τ is zero. The Liouville-von Neumann equation, which is a first-order differential equation with respect to time, describes a Markov process.

There is no rigorous prescription how to select the set of relevant observables $\{B_n\}$. The more relevant observables are selected so that their averages with $\rho_{\rm rel}(t)$ reproduce already the correctly known averages $\langle B_n \rangle^t$, see Eq. (18), the less the effort to produce the missing correlations dynamically, and the less relaxation time τ is needed. Taking into account that usually perturbation theory is used to treat the dynamical time evolution (23), a lower order of perturbation theory is then sufficient. We discuss this issue in Section 3.

In conclusion, the selection of the set of relevant observables is arbitrary, as a minimum the constants of motion C_n have to be included because their relaxation time is infinite, their averages cannot be produced dynamically. The resulting $\rho_{\rm NSO}(t)$ (26) should not depend on the (arbitrary) choice of relevant observables $\{B_n\}$ if the limit $\varepsilon \to 0$ is correctly performed. However, usually perturbation theory is applied, so that the result will depend on the selection of the set of relevant observables. The inclusion of long-living correlations into $\{B_n\}$ allows to use lower order perturbation expansions to obtain acceptable results.

2.3.3 Entropy of the nonequilibrium state

An intricate problem is the definition of entropy for the nonequilibrium state. In nonequilibrium, entropy is produced, as investigated in the phenomenological approach to the thermodynamics of irreversible processes, considering currents induced by the generalized forces.

Such a behavior occurs for the relevant entropy defined by the relevant distribution (20),

$$S_{\rm rel}(t) = -k_{\rm B}{\rm Tr}\{\rho_{\rm rel}(t) \ln \rho_{\rm rel}(t)\}. \tag{29}$$

A famous example that shows the increase of the relevant entropy with time is the Boltzmann H theorem where the relevant observables to define the nonequilibrium state are the occupation numbers of the single-particle states, i.e., the distribution function, see Section 3.2 for discussion.

Note that the increase of entropy cannot be solved this way. It is related to so-called coarse graining. The information about the state is reduced because the degrees of freedom to describe the system are reduced. This may be an averaging in phase space over small cells. The loss of information then gives the increase of entropy. This procedure is artificial, anthropomorphic, depending on our way to describe the details of a process.

The method of nonequilibrium statistical operator $\rho_{\mathrm{NSO}}(t)$ allows to extend the set of relevant observables arbitrarily so that the choice of the set of relevant observables seems to be irrelevant. All missing correlations are produced dynamically. We can start with any set of relevant operators, but have to wait for a sufficient long time to get the correct statistical operator, or to go to very small ϵ. A possible definition of the entropy would be

$$S_{\mathrm{NSO}}(t) = -k_{\mathrm{B}}\mathrm{Tr}\{\rho_{\mathrm{NSO}}(t)\ln \rho_{\mathrm{NSO}}(t)\}. \tag{30}$$

The destruction of the reversibility of the von Neumann equation (27) is connected with the source term on the right-hand side that produces the mixing by averaging over the past in Eq. (25). This source term is responsible for the entropy production. At present, there is no proof that the entropy $S_{\mathrm{NSO}}(t)$ will increase also in the limit $\epsilon \to +0$.

3. Applications

The NSO method is a fundamental step in deriving equations of evolution to describe nonequilibrium phenomena. It can be shown that any currently used description can be deduced from this approach. We give three typical examples, the quantum master equations, see [9, 10], kinetic theory, see [11], and linear response theory, see [12]. In all of these applications, we have to define the set of relevant observables, and to eliminate the Lagrange parameters determined by the self-consistency conditions. We shortly outline these applications, for a more exhaustive presentation see [3–5].

3.1 Quantum master equation

3.1.1 Open systems

The main issue is that any physical system cannot be completely separated from the surroundings, so that the isolated system is only a limiting case of the open system which is in contact with a bath. More general, we subdivide the degrees of freedom of the total system into the relevant degrees of freedom which describe the system S under consideration, and the irrelevant part describing the bath B. Examples are a harmonic oscillator coupled to a bath consisting of harmonic oscillators, such as an oscillating molecule interacting with phonons or photons, or radiation from a single atom embedded in the bath consisting of photons, see below.

The Hamiltonian H of the open system can be decomposed

$$H = H_{\mathrm{S}} + H_{\mathrm{B}} + H_{\mathrm{int}}. \tag{31}$$

The system Hamiltonian acts only in the Hibert space of the system states leaving the bath states unchanged. It is expressed in terms of the system observables

A_ν. The bath Hamiltonian acts only in the Hilbert space of the bath states leaving the system states unchanged. It is expressed in terms of the bath observables B_μ. Both sets of operators are assumed to be hermitean and independent so that $[A_\nu, B_\mu] = 0$.

We project out the relevant part of the nonequilibrium statistical operator $\rho(t)$

$$\rho_s(t) = \mathrm{Tr}_\mathrm{B}\rho(t) \tag{32}$$

where the trace over the bath can be performed after the eigenstates of the bath are introduced. The operator Tr_B means the trace over the quantum states of the heat bath. If we have no further information, we construct the relevant statistical operator taking the equilibrium distribution $\rho_\mathrm{B} = \rho_\mathrm{eq}$ (13) for the irrelevant degrees of freedom,

$$\rho_\mathrm{rel}(t) = \rho_s(t)\rho_\mathrm{B}. \tag{33}$$

3.1.2 Born-Markov approximation

Starting with the extended Liouville-von Neumann equation (27), we perform the trace Tr_B over the variables of the bath (see Eq. (32)),

$$\frac{\partial}{\partial t}\rho_s(t) - \frac{1}{\mathrm{i}\hbar}[H_s, \rho_s(t)] = \frac{1}{\mathrm{i}\hbar}\mathrm{Tr}_\mathrm{B}[H_\mathrm{int}, \rho(t)] \tag{34}$$

since the remaining terms disappear and $\frac{1}{\mathrm{i}\hbar}\mathrm{Tr}_\mathrm{B}(H_\mathrm{B}\rho(t) - \rho(t)H_\mathrm{B}) = 0$ because of cyclic invariance of the trace Tr_B. To obtain a closed equation for $\rho_s(t)$, the full nonequilibrium statistical operator $\rho(t)$ occurring on the right-hand side has to be eliminated.

For this, we calculate the time evolution of the irrelevant part of the statistical operator $\Delta\rho(t) = \rho(t) - \rho_{rel}(t)$,

$$\frac{\partial}{\partial t}\Delta\rho(t) = \frac{\partial}{\partial t}\rho(t) - \left[\frac{\partial}{\partial t}\rho_s(t)\right]\rho_\mathrm{B} \tag{35}$$

inserting the time evolution for $\rho(t)$ (8) and $\rho_s(t)$ (34) given above:

$$\left(\frac{\partial}{\partial t} + \varepsilon\right)\Delta\rho(t) = \frac{1}{\mathrm{i}\hbar}[H, \rho(t)] - \frac{1}{\mathrm{i}\hbar}[H_s, \rho_s(t)]\rho_\mathrm{B} - \rho_\mathrm{B}\frac{1}{\mathrm{i}\hbar}\mathrm{Tr}_\mathrm{B}[H_\mathrm{int}, \rho(t)]. \tag{36}$$

We eliminate $\rho(t) = \Delta\rho(t) + \rho_s(t)\rho_\mathrm{B}$ and collect all terms with $\Delta\rho(t)$ on the left-hand side. We can assume that $\langle H_\mathrm{int}\rangle_\mathrm{B} = \mathrm{Tr}_\mathrm{B}(H_\mathrm{int}\rho_\mathrm{B}) = 0$ because the heat bath do not exert external forces on the system (if not, replace H_s by $H_s + \langle H_\mathrm{int}\rangle_\mathrm{B}$ and H_int by $H_\mathrm{int} - \langle H_\mathrm{int}\rangle_\mathrm{B}$) so that also $(\mathrm{Tr}_\mathrm{B}[H_\mathrm{int}, \rho_\mathrm{B}])\rho_s(t) = 0$ and the last term $-\rho_\mathrm{B}\frac{1}{\mathrm{i}\hbar}\mathrm{Tr}_\mathrm{B}\{H_\mathrm{int}\rho_\mathrm{B}\}\rho_s(t) + \rho_\mathrm{B}\rho_s(t)\frac{1}{\mathrm{i}\hbar}\mathrm{Tr}_\mathrm{B}\{\rho_\mathrm{B}H_\mathrm{int}\}$ vanishes. Also, the term $\frac{1}{\mathrm{i}\hbar}[H_\mathrm{B}, \rho_s(t)\rho_\mathrm{B}]$ disappears since $[H_\mathrm{B}, \rho_\mathrm{B}] = 0$.

We obtain

$$\left(\frac{\partial}{\partial t} + \varepsilon\right)\Delta\rho(t) - \frac{1}{\mathrm{i}\hbar}[(H_s + H_\mathrm{int} + H_\mathrm{B}), \Delta\rho(t)] + \rho_\mathrm{B}\frac{1}{\mathrm{i}\hbar}\mathrm{Tr}_\mathrm{B}[H_\mathrm{int}, \Delta\rho(t)] = \frac{1}{\mathrm{i}\hbar}[H_\mathrm{int}, \rho_s(t)\rho_\mathrm{B}]. \tag{37}$$

The deviation $\Delta\rho(t)$ vanishes when $H_\mathrm{int} \to 0$. In lowest order with respect to H_int, the solution is found as

$$\Delta\rho(t) = \int_{-\infty}^{t} dt'\, e^{-\varepsilon(t-t')} e^{\frac{1}{i\hbar}(t-t')(H_s+H_B)} \frac{1}{i\hbar}[H_{\text{int}}, \rho_s(t')\rho_B] e^{-\frac{1}{i\hbar}(t-t')(H_s+H_B)}. \quad (38)$$

Inserting the solution (38) into the equation of motion of $\rho_s(t)$ (34), a closed equation of evolution is obtained eliminating $\rho(t)$. In the lowest (second) order with respect to the interaction considered here, memory effects are neglected. We can use the unperturbed dynamics to replace $\rho_s(t') = e^{-\frac{1}{i\hbar}(t-t')H_s}\rho_s(t)e^{\frac{1}{i\hbar}(t-t')H_s}$ and $H_{\text{int}}(\tau) = e^{-\frac{1}{i\hbar}\tau(H_s+H_B)}H_{\text{int}}e^{\frac{1}{i\hbar}\tau(H_s+H_B)}$ so that after a shift of the integration variable

$$\frac{\partial}{\partial t}\rho_s(t) - \frac{1}{i\hbar}[H_s, \rho_s(t)] = -\frac{1}{\hbar^2}\int_{-\infty}^{0} d\tau\, e^{\varepsilon\tau}\,\text{Tr}_B[H_{\text{int}}, [H_{\text{int}}(\tau), \rho_s(t)\rho_B]] = D[\rho_s(t)]. \quad (39)$$

This result is described as quantum master equation in Born approximation. For higher orders of H_{int} see [4, 5].

3.1.3 Rotating wave approximation and Lindblad form

We assume that the interaction has the form

$$H_{\text{int}} = \sum_\alpha A_\alpha \otimes B_\alpha. \quad (40)$$

We use the interaction picture that coincides at t_0 with the Schrödinger picture,

$$O^{(\text{int})}(t-t_0) = e^{i(H_S+H_B)(t-t_0)/\hbar} O e^{-i(H_S+H_B)(t-t_0)/\hbar} \quad (41)$$

for any operator O. In particular, we denote

$$\begin{aligned} \mathcal{D}^{(\text{int})}[\rho_S(t)](t-t_0) &= e^{i(H_S+H_B)(t-t_0)/\hbar}\mathcal{D}[\rho_S(t)]e^{-i(H_S+H_B)(t-t_0)/\hbar}, \\ \rho_S^{(\text{int})}(t;t-t_0) &= e^{iH_S(t-t_0)/\hbar}\rho_S(t)e^{-iH_S(t-t_0)/\hbar} \end{aligned} \quad (42)$$

(note that H_B commutes with $\rho_S(t)$ which is defined in the Hilbert space $\mathcal{H}_S$). Then, the dynamical evolution of the system is given by

$$\frac{\partial}{\partial t}\rho_S^{(\text{int})}(t;t-t_0) = \mathcal{D}^{(\text{int})}[\rho_S(t)](t-t_0). \quad (43)$$

On the left-hand side, we cancel H_B because it commutes with the system variables. The right-hand side, the influence term, has the form (note that ρ_B commutes with H_B)

$$D^{(\text{int})}[\rho_S(t)](t-t_0) = -\frac{1}{\hbar^2}\int_{-\infty}^{t} dt'\, e^{-\varepsilon(t-t')}\text{Tr}_B\left[H_{\text{int}}^{(\text{int})}(t-t_0), \left[H_{\text{int}}^{(\text{int})}(t'-t_0), \rho_S^{(\text{int})}(t;t-t_0)\right]\right]\rho_B. \quad (44)$$

In zeroth order of interaction, $\rho_S^{(\text{int})}(t;t-t_0) = e^{iH_S(t-t_0)/\hbar}\rho_S(t)e^{-iH_S(t-t_0)/\hbar}$ is not depending on t because the derivative with respect to t vanishes. This fact has already been used when in the Markov approximation $\rho_S(t')$ is replaced by $\rho_S(t)$. This corresponds to the Heisenberg picture where the state of the system does not change with time. The time dependence of averages is attributed to the temporal changes of the observables.

To include the interaction, we characterize the dynamics of the system observable A introducing the spectral decomposition with respect to the (discrete) eigenstates $|\phi_n\rangle$ of H_S. We introduce the eigen-energies E_n^s of the system S according to $H_S|\phi_n\rangle = E_n^s|\phi_n\rangle$, and with $\int_{-\infty}^{\infty} \exp[ikx]dx = 2\pi\delta(k)$,

$$\begin{aligned} A(\omega) &= \int_{-\infty}^{\infty} \mathrm{d}t e^{i\omega(t-t_0)} e^{iH_S(t-t_0)/\hbar} A e^{-iH_S(t-t_0)/\hbar} = A^\dagger(-\omega) \\ &= 2\pi\hbar \sum_{nm} |\phi_n\rangle\langle\phi_n|A|\phi_m\rangle\langle\phi_m|\delta(E_n^s - E_m^s + \hbar\omega) \end{aligned} \tag{45}$$

(the index α in (40) is dropped). In interaction picture (A commutes with the bath observables) we have $e^{iH_S(t-t_0)/\hbar}Ae^{-iH_S(t-t_0)/\hbar} = \int_{-\infty}^{\infty} \mathrm{d}\omega/(2\pi) \exp[-i\omega(t-t_0)] \times A(\omega)$. Now, we find for the influence term

$$\mathcal{D}^{(\mathrm{int})}[\rho_S(t)](t-t_0) = -\frac{1}{\hbar^2}\int_{-\infty}^{t} \mathrm{d}t' \int_{-\infty}^{\infty} \frac{\mathrm{d}\omega}{2\pi} \int_{-\infty}^{\infty} \frac{\mathrm{d}\omega'}{2\pi} e^{\varepsilon(t'-t)} e^{-i\omega'(t'-t)} e^{-i(\omega+\omega')(t-t_0)}$$
$$\times\left\{\langle \mathrm{B}(t'-t)\mathrm{B}\rangle_\mathrm{B}[A(\omega), \rho_S^{(\mathrm{int})}(t;t-t_0)A(\omega')] + \langle BB(t'-t)\rangle_\mathrm{B}[A(\omega')\rho_S^{(\mathrm{int})}(t;t-t_0), A(\omega)]\right\} \tag{46}$$

with the time-dependent bath operators $\mathrm{B}(t'-t) = \exp[iH_\mathrm{B}(t'-t)/\hbar]\mathrm{B} \exp[-iH_\mathrm{B}(t'-t)/\hbar]$.

We can perform the integral over t' that concerns the bath observables. The bath enters via equilibrium auto-correlation functions of the time-dependent bath operators $\mathrm{B}_\alpha(\tau)$. We introduce the Laplace transform of the bath correlation function (the response function of the bath)

$$\Gamma(\omega) = \int_0^{\infty} d\tau e^{i(\omega+i\epsilon)\tau/\hbar} \mathrm{Tr}_\mathrm{B}\{\rho_\mathrm{B}\mathrm{B}^\dagger(\tau)\mathrm{B}\} = \frac{1}{2}\gamma(\omega) + \mathrm{i}\frac{1}{\hbar}S(\omega) \tag{47}$$

that is a matrix $\Gamma_{\alpha\beta}(\omega)$ if the observable B has several components. We find in short notation

$$\begin{aligned} &\mathcal{D}^{(\mathrm{int})}[\rho_S(t)](t-t_0) = -\int_{-\infty}^{\infty} \frac{\mathrm{d}\omega}{2\pi} \int_{-\infty}^{\infty} \frac{\mathrm{d}\omega''}{2\pi} e^{i(\omega''-\omega)(t-t_0)} \\ &\times\left\{\Gamma_2(\omega'')\left[A(\omega), \rho_S^{(\mathrm{int})}(t;t-t_0)A^\dagger(\omega'')\right] + \Gamma_1(\omega'')\left[A^\dagger(\omega'')\rho_S^{(\mathrm{int})}(t;t-t_0), A(\omega)\right]\right\} \end{aligned} \tag{48}$$

after the transformation $\omega' \to -\omega''$ and using Eq. (45). Note that this expression for the influence term is real because the second contribution is the Hermitean conjugated of the first contribution. Using symmetry properties, all correlation functions of bath variables are related to $\Gamma(\omega)$.

The expression $\rho_S^{(\mathrm{int})}(t;t-t_0) = e^{iH_S(t-t_0)/\hbar}\rho_S(t)e^{-iH_S(t-t_0)/\hbar}$ is not depending on time t because in the Heisenberg picture (we consider the lowest order of interaction) the state of the system does not depend on time. Oscillations with $e^{i(\omega-\omega'')(t-t_0)}$ occur that vanish for $\omega'' = \omega$. The rotating wave approximation (RWA) takes into account only contributions with $\omega'' = \omega$ that are not depending on t_0. Oscillations with $e^{i(\omega-\omega')(t-t_0)}$, $\omega' - \omega \neq 0$ exhibit a phase, depending on t_0. Any process of dephasing will damp down these oscillations.

In the case of a discrete spectrum, the spectral function (45) can be used, and the integrals over ω, ω'' are replaced by sums over the eigenstates $|\phi_n\rangle$ of the system S:

$$\mathcal{D}^{(\text{int})}[\rho_S(t)](t-t_0) = -\frac{1}{\hbar^2}\sum_{nn',\,mm'} e^{\left(E^s_n - E^s_{n'} - E^s_m + E^s_{m'}\right)(t-t_0)/\hbar}\Gamma\left(\left(E^s_n - E^s_m\right)/\hbar\right)$$
$$\times\Big[|\phi_n\rangle\langle\phi_n|A|\phi_m\rangle\langle\phi_m|e^{\left(E^s_m - E^s_{m'}\right)(t-t_0)/\hbar}\rho_S(t)|\phi_{m'}\rangle\langle\phi_{m'}|A|\phi_{n'}\rangle\langle\phi_{n'}|$$
$$-|\phi_{m'}\rangle\langle\phi_{m'}|A|\phi_{n'}\rangle\langle\phi_{n'}|\phi_n\rangle\langle\phi_n|A|\phi_m\rangle\langle\phi_m|e^{\left(E^s_m - E^s_{m'}\right)(t-t_0)/\hbar}\rho_S(t)\Big] + \text{h.c.} \tag{49}$$

The rotating wave approximation means that $n = n'$, $m = m'$ so that

$$\mathcal{D}^{(\text{int})}[\rho_S(t)](t-t_0) = -\frac{1}{\hbar^2}\sum_{n,\,m}\Gamma\left(\left(E^s_n - E^s_m\right)/\hbar\right)$$
$$\times[|\phi_n\rangle\langle\phi_n|A|\phi_m\rangle\langle\phi_m|\rho_S(t)|\phi_m\rangle\langle\phi_m|A|\phi_n\rangle\langle\phi_n| - |\phi_m\rangle\langle\phi_m|A|\phi_n\rangle\langle\phi_n|A|\phi_m\rangle\langle\phi_m|\rho_S(t)] + \text{h.c.} \tag{50}$$

The generalization to a more complex coupling to a bath (40) is straightforward, leading to matrices. More difficult is the discussion if the spectral function $A(\omega)$ is continuous, see [5]. Going back to the Schrödinger picture we have

$$\mathcal{D}[\rho_S(t)] = \int d\omega \sum_{\alpha\beta}\Gamma_{\alpha\beta}(\omega)\left[A_\beta(\omega)\rho_S(t)A^\dagger_\alpha(\omega) - A^\dagger_\alpha(\omega)A_\beta(\omega)\rho_S(t)\right] + \text{h.c.} \tag{51}$$

The influence term $D[\rho_S(t)]$ cannot be given in the form of a commutator of an effective Hamiltonian with $\rho_S(t)$ that characterizes the Hamiltonian dynamics. Only a part can be separated that contributes to the reversible Hamiltonian dynamics, whereas the remaining part describes irreversible evolution in time and is denoted as dissipator $D'[\rho_S(t)]$.

With $\Gamma_{\alpha\beta}(\omega) = \gamma_{\alpha\beta}(\omega)/2 + iS_{\alpha\beta}(\omega)$, we introduce the Hermitian operator $H_{\text{infl}} = \int d\omega \sum_{\alpha\beta} S_{\alpha\beta}(\omega)A^\dagger_\alpha(\omega)A_\beta(\omega)$ and obtain the quantum master equation

$$\frac{\partial}{\partial t}\rho_S(t) - \frac{1}{i\hbar}[H_S, \rho_S(t)] - \frac{1}{i\hbar}[H_{\text{infl}}, \rho_S(t)] = D'[\rho_S(t)]. \tag{52}$$

The dissipator has the form

$$\mathcal{D}'[\rho_S(t)] = \int d\omega \sum_{\alpha\beta}\gamma_{\alpha\beta}(\omega)\left[A_\beta(\omega)\rho_S(t)A^\dagger_\alpha(\omega) - \frac{1}{2}\left\{A^\dagger_\alpha(\omega)A_\beta(\omega), \rho_S(t)\right\}\right] \tag{53}$$

where $\{A, B\} = AB + BA$ denotes the anticommutator. The influence Hamiltonian H_{infl} commutes with the system Hamiltonian, $[H_S, H_{\text{infl}}] = 0$, because the operator $A^\dagger_\alpha(\omega)A_\beta(\omega)$ commutes with H_S. It is often called the Lamb shift Hamiltonian since it leads to a shift of the unperturbed energy levels influenced by the coupling of the system to the reservoir, similar to the Lamb shift in QED. The Lindblad form follows by diagonalization of the matrices $\gamma_{\alpha\beta}(\omega)$,

$$\mathcal{D}'[\rho_S(t)] = \sum_k \gamma_k \left[A_k \rho_S(t) A_k^\dagger - \frac{1}{2} \{ A_k^\dagger A_k, \rho_S(t) \} \right]. \tag{54}$$

3.1.4 Example: harmonic oscillator in a bath

A typical example is the absorption or emission of light. An isolated atom (e.g., hydrogen) is usually treated with the Schrödinger equation which gives the well-known energy eigenvalues and the corresponding eigenstates. However, this is not correct, and the finite (natural) linewidth indicate that the energetically sharp eigenstates have not an infinite life-time. The coupling to the environment, the electromagnetic field (even in the vacuum at $T = 0$) leads to transitions and a finite life-time. The electromagnetic field which is considered as bath can be represented as a system of harmonic oscillators (for each mode of the field), and the interaction with the atomic system is (dipole approximation, dipole moment $\mathbf{D} = e\mathbf{r}$)

$$H_{\text{int}} = -e\mathbf{r} \cdot \mathbf{E} = -\mathbf{D} \cdot \mathbf{E}. \tag{55}$$

We discuss this phenomenon of radiation in a simplified version [5]. We consider a one-dimensional harmonic oscillator with the eigen-frequency ω_0,

$$H_S = \frac{1}{2m} p^2 + \frac{m\omega_0^2}{2} x^2 = \hbar\omega_0 \left(a^\dagger a + \frac{1}{2} \right), \tag{56}$$

with the creation $a^\dagger = (m\omega_0/2\hbar)^{1/2} x - \mathrm{i}/(2\hbar m\omega_0)^{1/2} p$ and destruction operator $a = (m\omega_0/2\hbar)^{1/2} x + \mathrm{i}/(2\hbar m\omega_0)^{1/2} p$ ($[a, a^\dagger] = 1$). The discrete eigenstates $|\phi_n\rangle$ of H_S are the well-known harmonic oscillator states, with eigen-energies $E_n^s = \hbar\omega_0(n + 1/2)$. The matrix elements of the construction operators are $\langle \phi_n | a | \phi_{n'} \rangle = \sqrt{n} \delta_{n'-1,n}$ and its adjoint complex. In interaction picture, the equations of motion are $\mathrm{d}a^\dagger(t)/\mathrm{d}t = \mathrm{i}\omega_0 a^\dagger(t)$, $\mathrm{d}a(t)/\mathrm{d}t = -\mathrm{i}\omega_0 a(t)$. The spectral representation reads

$$a^\dagger(\omega) = 2\pi \sum_n \sqrt{n+1} |\phi_{n+1}\rangle\langle\phi_n| \delta(\omega + \omega_0), \qquad a(\omega) = 2\pi \sum_n \sqrt{n} |\phi_{n-1}\rangle\langle\phi_n| \delta(\omega - \omega_0). \tag{57}$$

At this moment, we do not specify the bath any more in detail. Suppose we have the solutions $|n\rangle$ of the energy eigenvalue problem $H_B|m\rangle = E_{B,m}|m\rangle$, then we can construct the statistical operator for the canonical distribution as

$$\rho^0_{B,mm'} = \langle m' | \rho_B | m \rangle = \delta_{mm'} \frac{1}{Z} \mathrm{e}^{-E_{B,m}/k_B T}, \qquad Z = \sum_m \mathrm{e}^{-E_{B,m}/k_B T}. \tag{58}$$

We introduce a weak coupling between the system and the bath

$$H_{\text{int}} = -exE = \lambda(a^\dagger + a)\mathrm{B}, \tag{59}$$

where the operator B acts only on the variables of the bath and commutes with a and $a^\dagger$. In interaction picture we have

$$H_{\text{int}}^{(\text{int})}(t - t_0) = \lambda \left(a^\dagger e^{\mathrm{i}\omega_0(t-t_0)} + a e^{-\mathrm{i}\omega_0(t-t_0)} \right) \mathrm{B}(t - t_0). \tag{60}$$

The influence term is calculated as given above. With the response function of the bath $\Gamma(\omega)$ (47) we find

$$\frac{\partial}{\partial t}\rho_S(t) - \frac{1}{i\hbar}[H_S, \rho_S(t)] - \frac{1}{i\hbar}[(S(\omega_0)a^\dagger a + S(-\omega_0)aa^\dagger, \rho_S(t)]$$
$$= \gamma(\omega_0)\left(a\rho_S(t)a^\dagger - \frac{1}{2}\{a^\dagger a, \rho_S(t)\}\right) + \gamma(-\omega_0)\left(a^\dagger\rho_S(t)a - \frac{1}{2}\{aa^\dagger, \rho_S(t)\}\right). \tag{61}$$

The curly brackets in the dissipator denote the anticommutator. There are eight additional terms containing aa or $a^\dagger a^\dagger$. In interaction picture, they are proportional to $e^{\pm 2i\omega_0(t-t_0)}$ and are dropped within the rotating wave approximation. For a bath in thermal equilibrium, using eigenstates the detailed balance relation is easily proven,

$$\gamma(-\omega_0) = \gamma(\omega_0)e^{-\hbar\omega_0/k_B T}. \tag{62}$$

The evolution equations for the averages $\langle a^\dagger\rangle^t = \mathrm{Tr}_S\{\rho_S a^\dagger\}$, $\langle a^\dagger a\rangle^t = \mathrm{Tr}_S\{\rho_S a^\dagger a\}$ are immediately calculated as

$$\frac{d}{dt}\langle a^\dagger\rangle^t = \mathrm{Tr}_S\left\{\frac{\partial}{\partial t}\rho_S(t)a^\dagger\right\} = \left(i\omega_0' - \frac{1}{2}[\gamma(\omega_0) - \gamma(-\omega_0)]\right)\langle a^\dagger\rangle^t \tag{63}$$

with the renormalized frequency $\omega_0' = \omega_0 + [S(\omega_0) + S(-\omega_0)]/\hbar$. The solution is

$$\langle a^\dagger\rangle^t = \langle a^\dagger\rangle^{t_0} e^{[i\omega_{0'} - \gamma(\omega_0)/2 + \gamma(-\omega_0)/2](t-t_0)}. \tag{64}$$

Similar expressions are obtained for $\langle a\rangle^t$. We find for the occupation number $\langle n\rangle^t = \langle a^\dagger a\rangle^t = p_n(t)$

$$\frac{d}{dt}\langle a^\dagger a\rangle^t = \gamma(-\omega_0) - [\gamma(\omega_0) - \gamma(-\omega_0)]\langle a^\dagger a\rangle^t \tag{65}$$

with the solution

$$\langle a^\dagger a\rangle^t = \langle a^\dagger a\rangle^{t_0} e^{-[\gamma(\omega_0)-\gamma(-\omega_0)](t-t_0)} + \frac{\gamma(-\omega_0)}{\gamma(\omega_0) - \gamma(-\omega_0)}\left[1 - e^{-[\gamma(\omega_0)-\gamma(-\omega_0)](t-t_0)}\right]. \tag{66}$$

The asymptotic behavior $t - t_0 \to \infty$ is determined by the properties of the bath,

$$\frac{\gamma(-\omega_0)}{\gamma(\omega_0) - \gamma(\omega_0)} = \frac{1}{e^{-\hbar\omega_0/k_B T} - 1} = n_B(\omega_0), \tag{67}$$

the system relaxes to the thermal equilibrium distribution that is independent on the initial distribution $\langle a^\dagger a\rangle^{t_0}$.

3.1.5 Electromagnetic field

As example for the response function of the bath, we give the result for the blackbody radiation (Maxwell field)

$$\Gamma_{ij}(\omega) = \int_0^\infty d\tau e^{i(\omega+i\epsilon)\tau}\langle E_i(\tau)E_j(0)\rangle_B = \delta_{ij}\left(\frac{1}{2}\gamma(\omega) + iS(\omega)\right) \tag{68}$$

with

$$\gamma(\omega) = \frac{4\omega^3}{3\hbar c^3}[1 + n_\mathrm{B}(\omega)], \quad S(\omega) = \frac{2}{3\pi\hbar c^3} P \int_0^\infty d\omega_k \omega_k^3 \left[\frac{1 + n_\mathrm{B}(\omega_k)}{\omega - \omega_k} + \frac{n_\mathrm{B}(\omega_k)}{\omega + \omega_k}\right]. \tag{69}$$

Note that the Planck distribution satisfies $n_\mathrm{B}(-\omega) = -[1 + n_\mathrm{B}(\omega)]$ such that $\gamma(\omega) = 4\omega^3[1 + n_\mathrm{B}(\omega)]/(3\hbar c^3)$ for $\omega > 0$ and $\gamma(\omega) = 4|\omega|^3 n_\mathrm{B}(|\omega|)/(3\hbar c^3)$ for $\omega < 0$.

The resulting quantum optical master equation which, e.g., describes the coupling of atoms to the radiation field $H_\mathrm{int} = -\mathbf{D} \cdot \mathbf{E}$ in dipole approximation,

$$\frac{\partial}{\partial t}\rho_\mathrm{S}(t) - \frac{1}{\mathrm{i}\hbar}[H_\mathrm{S}, \rho_\mathrm{S}(t)] - \frac{1}{\mathrm{i}\hbar}[H_\mathrm{infl}, \rho_\mathrm{S}(t)] = \mathcal{D}'[\rho_\mathrm{S}(t)], \tag{70}$$

has the Lindblad form. The influence Hamiltonian $H_\mathrm{infl} = \int d\omega \hbar S(\omega) \mathbf{D}^\dagger(\omega) \cdot \mathbf{D}(\omega)$ leads to a renormalization of the system Hamiltonian H_S that is induced by the vacuum fluctuations of the radiation field (Lamb shift) and by the thermally induced processes (Stark shift). The dissipator of the quantum master equation reads

$$\begin{aligned} \mathcal{D}'[\rho_\mathrm{S}(t)] = &\int_0^\infty d\omega \frac{4\omega^3}{3\hbar c^3}[1 + n_\mathrm{B}(\omega)] \left[\mathbf{D}(\omega)\rho_\mathrm{S}(t)\mathbf{D}^\dagger(\omega) - \frac{1}{2}\left\{\mathbf{D}^\dagger(\omega)\mathbf{D}(\omega), \rho_\mathrm{S}(t)\right\}\right] \\ &+ \int_0^\infty d\omega \frac{4\omega^3}{3\hbar c^3} n_\mathrm{B}(\omega) \left[\mathbf{D}^\dagger(\omega)\rho_\mathrm{S}(t)\mathbf{D}(\omega) - \frac{1}{2}\left\{\mathbf{D}(\omega)\mathbf{D}^\dagger(\omega), \rho_\mathrm{S}(t)\right\}\right], \end{aligned} \tag{71}$$

where the integral over the negative frequencies has been transformed into positive frequencies. This result can be interpreted in a simple way. The application of the destruction operator $\mathbf{D}(\omega)$ on a state of the system lowers its energy by the amount $\hbar\omega$ and describes the emission of a photon. The transition rate $\frac{4\omega^3}{3\hbar c^3}[1 + n_\mathrm{B}(\omega)]$ contains the spontaneous emission as well as the thermal emission of photons. The term $\mathbf{D}^\dagger(\omega)$ gives the creation of excitations with transition rate $\frac{4\omega^3}{3\hbar c^3} n_\mathrm{B}(\omega)$ describing the absorption of photons.

3.1.6 The Pauli equation

We consider a system whose state is described by the observable A, and which takes the value a. This can be a set of numbers in the classical case that describe the degrees of freedom we use as relevant variables. In the quantum case, this is a set of relevant observables that describe the state of the system. The eigenvalue a corresponds to a state vector $|a\rangle$ in the Hilbert space.

At time t, we expect a probability distribution $p_1(a, t)$ to find the system in state a, if the property A is measured. The change of the probability $p_1(a, t)$ with time is described by a master equation or balance equation

$$\frac{d}{dt}p_1(a, t) = \sum_{a' \neq a} \left[w_{aa'}p_1(a', t) - w_{a'a}p_1(a, t)\right]. \tag{72}$$

In the context of the time evolution of a physical system, this master equation is also denoted as Pauli equation. We derive it from a microscopical approach using perturbation theory. The statistical operator $\rho(t)$ follows the von Neumann equation of motion (8) with the Hamiltonian

$$H = H^0 + \lambda H' \tag{73}$$

where the solution of the eigenvalue problem for H^0 is known, $H^0|n\rangle = E_n|n\rangle$. The probabilities to find the system in the state $|n\rangle$ are given by the diagonal elements of $\rho(t)$ in this representation,

$$p_1(n,t) = \langle n|\rho(t)|n\rangle. \tag{74}$$

First, we consider the special case $\lambda = 0$, where the von Neumann equation is easily solved:

$$\rho_{nm}(t) = \langle n|\rho(t)|m\rangle = e^{-i\omega_{nm}(t-t_0)}\rho_{nm}(t_0), \qquad \hbar\omega_{nm} = E_n - E_m \tag{75}$$

if $\rho_{nm}(t_0)$ is given. The nondiagonal elements $\rho_{nm}(t)$, $n \neq m$ are oscillating. The periodic time dependence of the density matrix that arises in the nondiagonal elements has nothing to do with any time evolution or irreversibility. It expresses the coherences in the system. The diagonal elements

$$\rho_{nn}(t) = p_1(n,t) = \langle n|\rho(t)|n\rangle \tag{76}$$

do not change with time and can be considered as conserved quantities if $\lambda = 0$.

To find the initial distribution, we consider the probabilities as relevant observables that describe the nonequilibrium state at t_0. If there are no further information on coherence, the relevant statistical operator is diagonal,

$$\rho_{\rm rel}(t_0) = \sum_n p_1(n,t_0)|n\rangle\langle n| = \sum_n p_1(n,t_0)\mathcal{P}_n. \tag{77}$$

We introduced the projection operator $\mathcal{P}_n = |n\rangle\langle n|$. The solution is $\rho(t) = \rho_{\rm rel}(t_0)$. The case $\lambda = 0$ is a trivial case, nothing happens.

Now, we consider a small perturbation as expressed by the parameter λ. As before, we consider the probabilities as relevant observables that describe the system in nonequilibrium. We project the diagonal part of the statistical operator,

$$\rho_{\rm rel}(t) = {\rm diag}[\rho(t)] = {\rm D}_n\rho(t) = \sum_n \mathcal{P}_n\rho(t)\mathcal{P}_n. \tag{78}$$

The difference $\rho_{\rm irrel}(t) = \rho(t) - \rho_{\rm rel}(t) = (1 - {\rm D}_n)\rho$ is the irrelevant part of the full statistical operator.

The problem to obtain the time evolution of the probabilities $p_1(n,t)$ is solved if we find an equation of evolution for $\rho_{\rm rel}(t)$. We use the method of the nonequilibrium statistical operator and start with the extended von Neumann equation (27). For the projection, we obtain (${\rm D}_n$ is linear and commutes with $\partial/\partial t$)

$$\frac{\partial}{\partial t}\rho_{\rm rel}(t) = \frac{1}{{\rm i}\hbar}{\rm D}_n[\lambda H', \rho_{\rm irrel}(t)]. \tag{79}$$

We assumed that H^0 is diagonal with $\rho_{\rm rel}(t)$ so that the commutator vanishes. Furthermore, the diagonal elements of the commutator of a diagonal matrix with an arbitrary matrix disappear. For the irrelevant part we have

$$\frac{\partial}{\partial t}\rho_{\rm irrel}(t) + \epsilon\rho_{\rm irrel}(t) - \frac{1}{{\rm i}\hbar}(1 - {\rm D}_n)[H, \rho_{\rm irrel}(t)] = \frac{1}{{\rm i}\hbar}(1 - {\rm D}_n)[\lambda H', \rho_{\rm rel}(t)]. \tag{80}$$

On the right-hand side, we can drop the projector D_n. Its action disappears because $\rho_{\rm rel}$ is diagonal. It is seen that $\rho_{\rm irrel}(t)$ is of the order λ.

In the remaining projection $(1-\mathrm{D}_n)\left[H^0,\rho_{\rm irrel}(t)\right]+(1-\mathrm{D}_n)[H',\rho_{\rm irrel}(t)]$, the second contribution is of second order in λ and will be dropped here because we consider only the lowest order in λ ($\rho_{\rm irrel}(t)$ is also of the order λ). This is denoted as Born approximation. We have

$$\frac{\partial}{\partial t}\rho_{\rm irrel}(t)+\varepsilon\rho_{\rm irrel}(t)-\frac{1}{\mathrm{i}\hbar}\left[H^0,\rho_{\rm irrel}(t)\right]=\frac{1}{\mathrm{i}\hbar}[\lambda H',\rho_{\rm rel}(t)]. \tag{81}$$

The solution is simple by integration,

$$\rho_{\rm irrel}(t)=\frac{1}{\mathrm{i}\hbar}\int_{-\infty}^{t}e^{\varepsilon(t_1-t)}e^{\frac{i}{\hbar}H^0(t_1-t)}[\lambda H',\rho_{\rm rel}(t_1)]e^{-\frac{i}{\hbar}H^0(t_1-t)}dt_1. \tag{82}$$

The proof is given by insertion.

With this expression for $\rho_{\rm irrel}(t)$, we find a closed equation for $\rho_{\rm rel}(t)$,

$$\frac{\partial}{\partial t}\rho_{\rm rel}(t)=-\frac{\lambda^2}{\hbar^2}\mathrm{D}_n\int_{-\infty}^{t}e^{\varepsilon(t_1-t)}\left[H',e^{\frac{i}{\hbar}H^0(t_1-t)}[H',\rho_{\rm rel}(t_1)]e^{-\frac{i}{\hbar}H^0(t_1-t)}\right]dt_1. \tag{83}$$

This result describes a memory effect. The change of $\rho_{\rm rel}(t)$ is determined by the values $\rho_{\rm rel}(t_1)$ at all previous times $t_1\le t$. In the Markov approximation, we replace $\rho_{\rm rel}(t_1)$ by $\rho_{\rm rel}(t)$ so that memory effects are neglected. This is justified in the limit $\lambda\to 0$ because then $\rho_{\rm rel}(t)$ changes only slowly with time. Then

$$\frac{\partial}{\partial t}\rho_{\rm rel}(t)=-\frac{\lambda^2}{\hbar^2}\mathrm{D}_n\int_{-\infty}^{t}e^{\varepsilon(t_1-t)}\left[H',\left[e^{\frac{i}{\hbar}H^0(t_1-t)}H'e^{-\frac{i}{\hbar}H^0(t_1-t)},\rho_{\rm rel}(t)\right]\right]dt_1. \tag{84}$$

This expression has similar structure as the QME (39) an can be treated in the same way. The right-hand side $D\rho_{\rm rel}(t)$ is related to the dissipator after subtracting the Lamb shift contribution.

Explicit expressions for the time evolution of the density matrix are obtained by projection on the basis $|n\rangle$. With the matrix elements $\langle n|\rho_{\rm rel}(t)|m\rangle=\delta_{n,m}p_1(n,t)$ as well as $\langle n|H^0|m\rangle=\delta_{n,m}E_n$ and $\langle n|H'|m\rangle=H'_{nm}$ we have

$$\begin{aligned}\frac{d}{dt}p_1(n,t)=&-\frac{\lambda^2}{\hbar^2}\sum_m H'_{nm}H'_{mn}\left[p_1(n,t)-p_1(m,t)\right]\\&\times\int_{-\infty}^{t}e^{\varepsilon(t_1-t)}\left[e^{\frac{i}{\hbar}(E_m-E_n)(t_1-t)}+e^{-\frac{i}{\hbar}(E_m-E_n)(t_1-t)}\right]dt_1.\end{aligned} \tag{85}$$

Performing the integral over t_1, we find [with the Dirac identity $\lim_{\epsilon\to+0}\frac{1}{x+i\epsilon}\equiv\mathcal{P}\frac{1}{x}-i\pi\delta(x)$] the Pauli equation

$$\frac{d}{dt}p_1(n,t)=\sum_{n'\neq n}\left[w_{nn'}p_1(n',t)-w_{n'n}p_1(n,t)\right]. \tag{86}$$

The transition rates are given by Fermi's Golden rule,

$$w_{nm}=\lim_{\epsilon\to 0}\frac{\lambda^2}{\hbar^2}\left|H'_{nm}\right|^2\left(\frac{1}{i\omega_{nm}+\epsilon}+\frac{1}{-i\omega_{nm}+\epsilon}\right)=\frac{2\pi\lambda^2}{\hbar}\left|H'_{nm}\right|^2\delta(E_n-E_m). \tag{87}$$

3.1.7 Properties of the Pauli equation

The transition rate w_{nm} obeys the condition of *detailed balance*, $w_{mn} = w_{nm}$, the inverse transition has the same rate. This follows because H' is hermitean,

$$\langle n|H'|m\rangle = \langle m|H'^{+}|n\rangle^{*} = \langle m|H'|n\rangle^{*}. \tag{88}$$

An important property is that it describes *irreversible evolution* with time. For the relevant entropy $S_{rel}(t) = -k_B \sum_n p_1(n,t) \ln p_1(n,t)$ we find

$$\begin{aligned}\frac{dS_{\rm rel}(t)}{dt} &= -k_B \sum_n \sum_m w_{nm} \left[p_1(m,t) - p_1(n,t)\right] \ln\left[p_1(n,t)\right] - k_B \sum_n \frac{p_1(n,t)}{p_1(n,t)} \frac{\partial p_1(n,t)}{\partial t} \\ &= \frac{1}{2} k_B \sum_n \sum_m w_{nm} \left[p_1(n,t) - p_1(m,t)\right] \left[\ln\left[p_1(n,t)\right] - \ln\left[p_1(m,t)\right]\right] \geq 0.\end{aligned} \tag{89}$$

We used $\frac{d}{dt} \sum_n p_1(n,t) = \frac{d}{dt} 1 = 0$ and interchanged n with m in the half of the expression. Since $\ln x$ is a monotonic function of x, the relation $(x_1 - x_2)(\ln x_1 - \ln x_2) \geq 0$ holds. Considering states n, m where transitions are possible, equilibrium ($dS_{\rm rel}(t)/dt = 0$) occurs if $p_1(m,t) = p_1(n,t)$; else $S_{\rm rel}(t)$ increases with time. Equipartition corresponds to the microcanonical ensemble in equilibrium.

3.1.8 Example: transition rates

We consider transitions between eigenstates of H_0 owing to interaction. A typical case is the collisions expressed by $a^{\dagger}_{k_1} a^{\dagger}_{k_2} a_{k'_2} a_{k'_1}$ between the (momentum) eigenstates $|k\rangle$ of H_0. This is discussed in the following section on kinetic theory. Another example is minimal coupling known from QFT between a Dirac fermionic field (electron) and the Maxwell bosonic field (photons), with ($E_k = \hbar^2 k^2/2m$, $\omega_q = c|\mathbf{q}|$)

$$H_0 = \sum_k E_k a^{\dagger}_{\mathbf{k}} a_{\mathbf{k}} + \sum_q \hbar\omega_q b^{\dagger}_{\mathbf{q}} b_{\mathbf{q}} \tag{90}$$

(spin and polarization variables are not indicated separately), and the interaction

$$H_{\rm int} = \sum_{k,k',q} v(kk',q) a^{\dagger}_{k'} a_{\mathbf{k}} b^{\dagger}_{\mathbf{q}} + \text{h.c.} \tag{91}$$

The transition rates (87) are calculated between the initial state $|n\rangle = |\mathbf{k}\rangle$, energy $E_n = E_k$, and the final state $|m\rangle = |k', \mathbf{q}\rangle$, energy $E_m = E_{k'} + \hbar\omega_q$ for emission in the vacuum state. For absorption, the corresponding process can be given. For free particles $|k\rangle = |\mathbf{k}, \sigma\rangle$, the matrix element $v(\mathbf{k}, \sigma, \mathbf{k}', \sigma', \mathbf{q}) \propto \delta_{k'+q,k}$ must fulfill momentum conservation. Together with the conservation of energy in Eq. (87), the second-order transition rate vanishes. Only in fourth order, different contributions (Compton scattering, pair creation) are possible. If considering an radiating atom, the electrons are moving in orbits around the nucleus, $|k\rangle = |nlm, \sigma\rangle$. Momentum conservation is not required, and the standard expressions (Fermi's Golden rule) for absorption and emission of light by an atom are obtained. The corresponding rate equation (86) describes natural line width, detailed balance, and thermal equilibrium as stationary solution.

3.1.9 Conclusions

Quantum master equations and the Pauli equation are fundamental expressions to describe nonequilibrium phenomena, such as one-step processes of excitation and deexcitation, two-level systems, nuclear decay, chemical reactions, and also conductivity where electrons are scattered by ions, etc. A basic assumption is the subdivision into a system and a bath. In Born-Markov approximation, correlations between system and bath (back-reactions) are neglected. Projection to diagonal elements of the reduced density matrix or the Rotating wave approximation lead to irreversible equations of evolution (dissipator) as derived by Zwanzig, Lindblad, Kossakowski, and others. Further developments of the theory are, e.g., the Nakajima-Zwanzig equation or the Quantum Fokker-Planck equation [4]. A fundamental problem is the subdivision in relevant (system) and irrelevant (bath) degrees of freedom. If correlations between the system and bath become relevant, the corresponding degrees of freedom of the bath must be included in the set of system variables.

3.2 Kinetic theory

Historically, nonequilibrium statistical physics was first developed as the kinetic theory of gases [7] by Boltzmann. We start with classical systems to explain the problem to be solved in kinetic theory. The more general case of quantum systems contains no additional complications, but the concepts become more evident in the classical limit. We give results for both cases, the general quantum case and the classical limit. Reduced distribution functions are considered as the relevant observables. Closed equations of evolution are obtained describing irreversible processes.

3.2.1 The Liouville equation

The standard treatment of a classical dynamical system can be given in terms of the Hamilton canonical equations. In classical mechanics, we have generalized coordinates and canonic conjugated momenta describing the state of the system, e.g., a point in the $6N$-dimensional phase space (Γ-space) in the case of N point masses. The $6N$ degrees of freedom $\{\boldsymbol{r}_1, \boldsymbol{p}_1 ... \boldsymbol{r}_N, \boldsymbol{p}_N\}$ define the microstate of the system. The evolution of a particular system with time is given by a trajectory in the phase space. Depending on the initial conditions different trajectories are taken.

Within statistical physics, instead of a special system, an ensemble of identical systems is considered, consisting of the same constituents and described by the same Hamiltonian, but at different initial conditions (microstates), which are compatible with the values of a given set of relevant observables characterizing the macrostate of the system. The probability of the realization of a macrostate by a special microstate, i.e., a point in the $6N$-dimensional phase space (Γ-space), is given by the N-particle distribution function $f_N(\boldsymbol{r}_i, \boldsymbol{p}_i, t)$ which is normalized,

$$\int d\Gamma f_N(\boldsymbol{r}_i, \boldsymbol{p}_i, t) = 1; \qquad d\Gamma = \frac{\mathrm{d}^N \boldsymbol{r}\, \mathrm{d}^N \boldsymbol{p}}{N! h^{3N}} = \frac{\mathrm{d}^{3N} x\, \mathrm{d}^{3N} p}{N! h^{3N}}. \tag{92}$$

In nonequilibrium, the N-particle distribution function depends on the time t.

The macroscopic properties can be evaluated as averages of the microscopic quantities $a(\boldsymbol{r}_i, \boldsymbol{p}_i)$ with respect to the distribution function $f_N(\boldsymbol{r}_i, \boldsymbol{p}_i, t)$:

$$\langle A \rangle^t = \int d\Gamma a(\boldsymbol{r}_i, \boldsymbol{p}_i) f_N(\boldsymbol{r}_i, \boldsymbol{p}_i, t). \tag{93}$$

In addition to these so-called mechanical properties there exist also thermal properties, such as entropy, temperature, and chemical potential. Instead of a dynamical variable, they are related to the distribution function. For example, the equilibrium entropy is given by

$$S_{\text{eq}} = -k_{\text{B}} \int d\Gamma f_N(\boldsymbol{r}_i, \boldsymbol{p}_i, t) \ln f_N(\boldsymbol{r}_i, \boldsymbol{p}_i, t) \tag{94}$$

We derive an equation of motion for the distribution function $f_N(\boldsymbol{r}_i, \boldsymbol{p}_i, t)$, the Liouville equation, see [5]:

$$\frac{\mathrm{d}f_N}{\mathrm{d}t} = \frac{\partial f_N}{\partial t} + \sum_{i=1}^{N} \left[\frac{\partial f_N}{\partial \boldsymbol{r}_i} \dot{\boldsymbol{r}}_i + \frac{\partial f_N}{\partial \boldsymbol{p}_i} \dot{\boldsymbol{p}}_i \right] = 0. \tag{95}$$

We shortly remember the quantum case. Instead of the N-particle distribution function $f_N(t)$, the statistical operator $\rho(t)$ is used to indicate the probability of a microstate in a given macrostate. The equation of motion is the von Neumann equation (8). Both equations are closely related and denoted as Liouville-von Neumann equation.

3.2.2 Classical reduced distribution functions

To evaluate averages, instead of the N-particle distribution function $f_N(\boldsymbol{r}_1, \ldots, \boldsymbol{r}_N; \boldsymbol{p}_1, \ldots, \boldsymbol{p}_N; t)$ often reduced s-particle distribution functions

$$f_s(\boldsymbol{r}_1, \ldots, \boldsymbol{p}_s; t) = \int \frac{\mathrm{d}^3\boldsymbol{r}_{s+1} \ldots \mathrm{d}^3\boldsymbol{p}_N}{(N-s)! h^{3(N-s)}} f_N(\boldsymbol{r}_1, \ldots, \boldsymbol{p}_N; t) \tag{96}$$

are sufficient. Examples are the particle density, the Maxwell distribution of the particle velocities, and the pair correlation function.

We are interested in the equations of motion for the reduced distribution functions. For classical systems, one finds a hierarchy of equations. From the Liouville equation, Eq. (95) without external potential,

$$\frac{\mathrm{d}f_N}{\mathrm{d}t} = \frac{\partial f_N}{\partial t} + \sum_i^N \boldsymbol{v}_i \frac{\partial f_N}{\partial \boldsymbol{r}_i} - \sum_{i \neq j}^N \frac{\partial V_{ij}}{\partial \boldsymbol{r}_i} \frac{\partial f_N}{\partial \boldsymbol{p}_i} = 0 \tag{97}$$

we obtain the equation of motion for the reduced distribution function f_s through integration over the $3(N-s)$ other variables:

$$\frac{\mathrm{d}f_s}{\mathrm{d}t} = \frac{\partial f_s}{\partial t} + \sum_{i=1}^{s} \boldsymbol{v}_i \frac{\partial f_s}{\partial \boldsymbol{r}_i} - \sum_{i \neq j}^{s} \frac{\partial V_{ij}}{\partial \boldsymbol{r}_i} \frac{\partial f_s}{\partial \boldsymbol{p}_i} = \sum_{i=1}^{s} \int \frac{\mathrm{d}^3\boldsymbol{r}_{s+1} \mathrm{d}^3\boldsymbol{p}_{s+1}}{h^3} \frac{\partial V_{i,s+1}}{\partial \boldsymbol{r}_i} \frac{\partial f_{s+1}(\boldsymbol{r}_1 \ldots \boldsymbol{p}_{s+1}, t)}{\partial \boldsymbol{p}_i}. \tag{98}$$

This hierarchy of equations is called BBGKY hierarchy, standing for Bogoliubov, Born, Green, Kirkwood, and Young.

The equation of motion (98) for the reduced distribution function f_s is not closed because on the right-hand side the higher order distribution function f_{s+1} appears. In its turn, f_{s+1} obeys a similar equation that contains f_{s+2}, etc. This structure of a system of equations is denoted as hierarchy. To obtain a kinetic equation that is a closed equation for the reduced distribution function, one has to truncate the BBGKY hierarchy, expressing the higher order distribution function f_{s+1} by the lower order distribution functions $\{f_1, \ldots, f_s\}$.

3.2.3 Quantum statistical reduced distributions

In the quantum case, the distribution function f_N is replaced by the statistical operator ρ that describes the state of the system, and the equation of motion is the von Neumann equation (8). The quantum statistical reduced density matrix is defined as average over creation and annihilation operators,

$$\rho_s(\boldsymbol{r}_1, \ldots, \boldsymbol{r}'_s, t) = \mathrm{Tr}\left\{\rho(t)\psi^\dagger(\boldsymbol{r}_1)\ldots\psi^\dagger(\boldsymbol{r}_s)\psi(\boldsymbol{r}'_s)\ldots\psi(\boldsymbol{r}'_1)\right\}. \tag{99}$$

It is related to correlation functions, the Wigner function, Green functions, dynamical structure factor, and others.

We consider the equations of motion for reduced distribution functions. For the single-particle density matrix in momentum representation, we have

$$\rho_1(\boldsymbol{p}, \boldsymbol{p}', t) = \mathrm{Tr}\{\rho(t)\psi^\dagger(\boldsymbol{p})\psi(\boldsymbol{p}')\}. \tag{100}$$

Derivation with respect to time gives

$$\frac{\partial}{\partial t}\rho_1(\boldsymbol{p}, \boldsymbol{p}', t) = \frac{1}{\mathrm{i}\hbar}\mathrm{Tr}\{[H, \rho]\psi^\dagger(\boldsymbol{p})\psi(\boldsymbol{p}')\} = \frac{1}{\mathrm{i}\hbar}\mathrm{Tr}\{\rho[\psi^\dagger(\boldsymbol{p})\psi(\boldsymbol{p}'), H]\}. \tag{101}$$

Similar as for the BBGKY hierarchy, we obtain in general a hierarchy of equations of the form

$$\frac{\partial\rho_s(t)}{\partial t} = \text{function of } \{\rho_s(t), \rho_{s+1}(t)\}. \tag{102}$$

Like in the classical case, we have to truncate this chain of equations. For example, in the Boltzmann equation for $f_1(t)$, the higher order distribution function $f_2(t)$ is replaced by a product of single-particle distribution functions $f_1(t)$.

3.2.4 Stoßzahlansatz and Boltzmann equation

To evaluate the averages of single-particle properties such as particle current or kinetic energy, only the single-particle distribution must be known. Then, the single-particle distribution contains the relevant information, the higher distributions are irrelevant and will be integrated over.

We are looking for an equation of motion for the single-particle distribution function $f_1(\boldsymbol{r}, \boldsymbol{p}, t)$, taking into account short range interactions and binary collisions. For the total derivative with respect to time we find, see Eq. (95),

$$\frac{\mathrm{d}f_1}{\mathrm{d}t} = \frac{\partial}{\partial t}f_1 + \dot{\boldsymbol{r}}\frac{\partial}{\partial \boldsymbol{r}}f_1 + \dot{\boldsymbol{p}}\frac{\partial}{\partial \boldsymbol{p}}f_1 = \frac{\partial}{\partial t}f_1 + \boldsymbol{v}\frac{\partial}{\partial \boldsymbol{r}}f_1 + \boldsymbol{F}\frac{\partial}{\partial \boldsymbol{p}}f_1 = 0.$$

The crucial point in this equation is the force $\boldsymbol{F}$. It is the sum of external forces $\boldsymbol{F}^{\text{ext}}$ acting on the system under consideration and all forces resulting from the interaction $V_{ij}(\boldsymbol{r}_i, \boldsymbol{r}_j)$ between the constituents of the system.

Before discussing the derivation of kinetic equations using the method of the nonequilibrium statistical operator, we give a phenomenological approach using empirical arguments. To describe the change in the distribution function f_1 due to collisions among particles, we write

$$\frac{\partial}{\partial t} f_1 = \left(\frac{\partial}{\partial t} f_1\right)_{\text{D}} + \left(\frac{\partial}{\partial t} f_1\right)_{\text{St}}, \tag{103}$$

where the drift term contains the external force,

$$\left(\frac{\partial}{\partial t} f_1\right)_{\text{D}} = -\boldsymbol{v}\frac{\partial}{\partial \boldsymbol{r}} f_1 - \boldsymbol{F}^{\text{ext}} \frac{\partial}{\partial \boldsymbol{p}} f_1 \tag{104}$$

and the internal interactions are contained in the collision term $\left(\frac{\partial}{\partial t} f_1\right)_{\text{St}}$ for which, from the BBGKY hierarchy (98), an exact expression has already been given:

$$\left(\frac{\partial}{\partial t} f_1\right)_{\text{St}} = \int \frac{\mathrm{d}^3 r' \mathrm{d}^3 p'}{h^3} \frac{\partial V(\boldsymbol{r}, \boldsymbol{r}')}{\partial \boldsymbol{r}} \frac{\partial}{\partial \boldsymbol{p}} f_2(\boldsymbol{rp}, \boldsymbol{r}'\boldsymbol{p}', t). \tag{105}$$

Collisions or interactions among particles occur due to the interaction potential $V(\boldsymbol{r}, \boldsymbol{r}')$, which depends on the coordinates of the two colliding partners. For every particle, one has to sum over collision with all partners in the system. In this way, we have an equation for the single-particle distribution function, but it is not closed because the right-hand side contains the two-particle distribution function $f_2(\boldsymbol{rp}, \boldsymbol{r}'\boldsymbol{p}', t)$.

As an approximation, similar to the master equation, we assume a balance between gain and loss:

$$\left(\frac{\partial f_1}{\partial t}\right)_{\text{St}} = G - L. \tag{106}$$

With some phenomenological considerations [5], we can find the collision term as

$$\left(\frac{\partial f_1}{\partial t}\right)_{\text{St}} = \int \mathrm{d}^3 \boldsymbol{v}_2 \int d\Omega \frac{\mathrm{d}\sigma}{d\Omega} |\boldsymbol{v}_1 - \boldsymbol{v}_2| \{ f_1(\boldsymbol{r}, \boldsymbol{v}_1', t) f_1(\boldsymbol{r}, \boldsymbol{v}_2', t) - f_1(\boldsymbol{r}, \boldsymbol{v}_1, t) f_1(\boldsymbol{r}, \boldsymbol{v}_2, t) \}, \tag{107}$$

where we have introduced the differential cross section

$$\frac{\mathrm{d}\sigma}{\mathrm{d}\Omega} = \frac{b(\vartheta)}{\sin\vartheta} \left| \frac{\mathrm{d}b(\vartheta)}{\mathrm{d}\vartheta} \right|. \tag{108}$$

Inserting expression (108) into Eq. (103), we obtain a kinetic equation only for the single-particle distribution, the Boltzmann equation.

3.2.5 Derivation of the Boltzmann equation from the nonequilibrium statistical operator

The relevant observable to describe the nonequilibrium state of the system is the single-particle distribution function. First, we consider *classical mechanics* where the single-particle distribution function is $f_1(\boldsymbol{r}, \boldsymbol{p}, t)$.

We can write the single-particle distribution as an average (93) of a microscopic (dynamic) variable, the single-particle density

$$f_1(\boldsymbol{r},\boldsymbol{p},t) = \langle n_1(\boldsymbol{r}_1,\dots,\boldsymbol{p}_N,\boldsymbol{r},\boldsymbol{p})\rangle^t, \quad n_1(\boldsymbol{r}_1,\boldsymbol{p}_1,\dots,\boldsymbol{r}_N,\boldsymbol{p}_N;\boldsymbol{r},\boldsymbol{p}) = \hbar^3 \sum_{i=1}^N \delta^3(\boldsymbol{r}-\boldsymbol{r}_i)\delta^3(\boldsymbol{p}-\boldsymbol{p}_i). \tag{109}$$

The self-consistency conditions (18) are realized with the Lagrange parameter $F_1(\boldsymbol{r},\boldsymbol{p},t)$. The relevant distribution $F_{\rm rel}$ reads (see (19) and replace $\sum_n$ by $\int d^3r d^3p/h^3$)

$$F_{\rm rel}(\boldsymbol{r}_1,\dots,\boldsymbol{p}_N,t) = \exp\left\{-\Phi(t) - \sum_{i=1}^N F_1(\boldsymbol{r}_i,\boldsymbol{p}_i,t)\right\}, \quad \Phi(t) = \ln\int \exp\left\{-\sum_{i=1}^N F_1(\boldsymbol{r}_i,\boldsymbol{p}_i,t)\right\} d\Gamma. \tag{110}$$

The constraints $f_1(\boldsymbol{r},\boldsymbol{p},t) \equiv \int F_{\rm rel}(\boldsymbol{r}_1,\dots,\boldsymbol{p}_N,t)\, n_1(\boldsymbol{r}_1,\dots,\boldsymbol{p}_N,\boldsymbol{r},\boldsymbol{p}) {\rm d}\Gamma$ are solved according to

$$f_1(\boldsymbol{r},\boldsymbol{p},t) = h^3 N\, {\rm e}^{-F_1(\boldsymbol{r},\boldsymbol{p},t)} \left\{\int {\rm e}^{-F_1(\boldsymbol{r},\boldsymbol{p},t)} {\rm d}^3\boldsymbol{r}\, {\rm d}^3\boldsymbol{p}\right\}^{-1}, \qquad F_1(\boldsymbol{r},\boldsymbol{p},t) = -\ln f_1(\boldsymbol{r},\boldsymbol{p},t). \tag{111}$$

This means, we can eliminate the Lagrange parameters $F_1(\boldsymbol{r},\boldsymbol{p},t)$ that are expressed in terms of the given distribution function $f_1(\boldsymbol{r},\boldsymbol{p},t)$. The relevant distribution is

$$F_{\rm rel}(\boldsymbol{r}_1,\dots,\boldsymbol{p}_N,t) = \frac{1}{Z_{\rm rel}} \prod_j f_1(\boldsymbol{r}_j,\boldsymbol{p}_j,t), \qquad Z_{\rm rel} = \int \prod_j f_1(\boldsymbol{r}_j,\boldsymbol{p}_j,t)\, {\rm d}\Gamma_N = \frac{N^N}{N!} \approx {\rm e}^N. \tag{112}$$

The *Boltzmann entropy* is then

$$S_{\rm rel}(t) = -k_{\rm B}\langle \ln F_{\rm rel}\rangle^t = -k_{\rm B}\int f_1(\boldsymbol{r},\boldsymbol{p},t) \ln\frac{f_1(\boldsymbol{r},\boldsymbol{p},t)}{\rm e} \frac{{\rm d}^3\boldsymbol{r}\, {\rm d}^3\boldsymbol{p}}{h^3}. \tag{113}$$

Below, we show that it increases with time for nonequilibrium distributions.

The relevant distribution can be used to derive the collision term (107), for details see [3]. We will switch over to the quantum case where the presentation is more transparent.

In the *quantum case*, we consider the single-particle density matrix. In the case of a homogeneous system ($n_1(\boldsymbol{r}) = n$), $\rho_1(\boldsymbol{p},\boldsymbol{p}')$ is diagonal. The set of relevant observables are the occupation number operators $\{n_{\boldsymbol{p}}\}$,

$$\langle n_{\boldsymbol{p}}\rangle^t = f_1(\boldsymbol{p},t). \tag{114}$$

Considering these mean values as given, we construct the relevant statistical operator as

$$\rho_{\rm rel}(t) = {\rm e}^{-\Phi(t) - \sum_p F_1(\boldsymbol{p},t) n_p}, \qquad \Phi(t) = \ln {\rm Tr}\, {\rm e}^{-\sum_p F_1(\boldsymbol{p},t) n_p}. \tag{115}$$

The Lagrange parameters $F_1(\boldsymbol{p},t)$ are obtained from the self-consistency conditions (114) similar to Eq. (111);

$$f_1(\boldsymbol{p},t) = \frac{\mathrm{Tr}\left\{\mathrm{e}^{-\sum_{p'} F_1(\boldsymbol{p}',t) n_{p'}} n_{\boldsymbol{p}}\right\}}{\mathrm{Tr}\left\{\mathrm{e}^{-\sum_{p'} F_1(\boldsymbol{p}',t) n_{p'}}\right\}} = \frac{\prod_i \sum_{n_i} \mathrm{e}^{-F_1(\boldsymbol{p}_i,t) n_i}\left(1+\delta_{\boldsymbol{p}_i,\boldsymbol{p}}(n_i-1)\right)}{\prod_i \sum_{n_i} \mathrm{e}^{-F_1(\boldsymbol{p}_i,t) n_i}} \tag{116}$$

so that

$$f_1(\boldsymbol{p},t) = \left\{\frac{1}{\mathrm{e}^{F_1(\boldsymbol{p},t)} \pm 1}, \begin{array}{ll} + & : \text{Fermions} \\ - & : \text{Bosons} \end{array}\right\}, \qquad F_1(\boldsymbol{p},t) = \ln\left[1 \mp f_1(\boldsymbol{p},t)\right] - \ln f_1(\boldsymbol{p},t). \tag{117}$$

As in the classical case, also in the quantum case, the Lagrange parameters can be eliminated explicitly.

We now *derive the Boltzmann equation for the quantum case*, see [3]. With the statistical operator (Eq. (25) after integration by parts)

$$\rho(t) = \rho_{\mathrm{rel}}(t) - \int_{-\infty}^{t} \mathrm{e}^{\epsilon(t_1-t)} \frac{\mathrm{d}}{\mathrm{d}t_1}\left\{\mathrm{e}^{-\frac{\mathrm{i}}{\hbar}H(t-t_1)}\rho_{\mathrm{rel}}(t_1)\mathrm{e}^{\frac{\mathrm{i}}{\hbar}H(t-t_1)}\right\}\mathrm{d}t_1, \tag{118}$$

With $\dot{n}_{\boldsymbol{p}} = \frac{\mathrm{i}}{\hbar}\left[H, n_{\boldsymbol{p}}\right]$, we get the time derivative of the single-particle distribution function

$$\frac{\partial}{\partial t} f_1(\boldsymbol{p},t) = \mathrm{Tr}\left\{\rho_{\mathrm{rel}}(t)\dot{n}_{\boldsymbol{p}}\right\} - \int_{-\infty}^{0} \mathrm{e}^{\epsilon t'}\,\mathrm{Tr}\left\{\dot{n}_{\boldsymbol{p}} \frac{\mathrm{d}}{\mathrm{d}t'}\left[\mathrm{e}^{\frac{\mathrm{i}}{\hbar}Ht'}\rho_{\mathrm{rel}}(t+t')\mathrm{e}^{-\frac{\mathrm{i}}{\hbar}Ht'}\right]\right\}\mathrm{d}t'. \tag{119}$$

Because the trace is invariant with respect to cyclic permutations and $\rho_{\mathrm{rel}}(t)$ commutes with $n_{\boldsymbol{p}}$, see (115),

$$\mathrm{Tr}\left\{\rho_{\mathrm{rel}}(t)\dot{n}_{\boldsymbol{p}}\right\} = \frac{\mathrm{i}}{\hbar}\mathrm{Tr}\left\{\rho_{\mathrm{rel}}\left[H,n_{\boldsymbol{p}}\right]\right\} = \frac{\mathrm{i}}{\hbar}\mathrm{Tr}\left\{H\left[n_{\boldsymbol{p}},\rho_{\mathrm{rel}}\right]\right\} = 0, \tag{120}$$

and Eq. (119) can be written as

$$\frac{\partial f_1}{\partial t} = \frac{1}{\hbar^2}\int_{-\infty}^{0} \mathrm{d}t'\,\mathrm{e}^{\epsilon t'}\,Tr\left\{\left[H,n_{\boldsymbol{p}}\right]\mathrm{e}^{\frac{\mathrm{i}}{\hbar}Ht'}\left[H,\rho_{\mathrm{rel}}\right]\mathrm{e}^{-\frac{\mathrm{i}}{\hbar}Ht'}\right\}, \tag{121}$$

if we neglect the explicit time dependence of $\rho_{\mathrm{rel}}(t)$ (no memory effects, the collision term is local in space and time). Next, we introduce two more integrations via delta functions to get rid of the time dependence in the trace:

$$\frac{\partial f_1}{\partial t} = \frac{1}{\hbar^2}\int_{-\infty}^{\infty}\mathrm{d}E\int_{-\infty}^{\infty}\mathrm{d}E'\int_{-\infty}^{0}\mathrm{d}t'\,\mathrm{e}^{\left[\epsilon+\frac{\mathrm{i}}{\hbar}(E-E')\right]t'}\,\mathrm{Tr}\left\{\left[V,n_{\boldsymbol{p}}\right]\delta(E-H)\left[V,\rho_{\mathrm{rel}}\right]\delta(E'-H)\right\}. \tag{122}$$

(We take into account that the kinetic energy in H commutes with $n_{\boldsymbol{p}}$ so that only the potential energy V remains.) This equation can be expressed by so-called T matrices, $T = V + V\frac{1}{E-H}T$,

$$\frac{\partial f_1}{\partial t} = \frac{\pi}{\hbar}\int dE\,\mathrm{Tr}\left\{[T, n_p]\delta(E - H^0)[T, \rho_{\rm rel}]\delta(E - H^0)\right\}, \tag{123}$$

For further treatment, we choose the approximation of binary collisions, that means that only two particles change their momentums during a collision. In second quantization, the T matrix is then

$$T \approx \sum_{p_1, p_2, p_{1'} p_{2'}} a^\dagger_{p_1} a^\dagger_{p_2} t(p_1, p_2, p'_1, p'_2) a_{p'_2} a_{p'_1} \delta(p_1 + p_2 - p'_1 - p'_2), \tag{124}$$

with the two-particle T matrix $t(p_1, p_2, p'_1, p'_2)$. With this T matrix, we find the collision term (time t is dropped)

$$\begin{aligned}\left(\frac{\partial f_1(p_1)}{\partial t}\right)_{\rm St} = & \sum_{p_2 p'_1 p'_2} w(p_1 p_2 p'_1 p_2)\{f_1(p'_1) f_1(p'_2)(1 \mp f_1(p_1))(1 \mp f_1(p_2)) \\ & - f_1(p_1) f_1(p_2)(1 \mp f_1(p'_1))(1 \mp f_1(p'_2))\}\end{aligned} \tag{125}$$

with the transition probability rate

$$w(p_1 p_2 p'_1 p'_2) = \frac{2\pi}{\hbar}|t(p_1 p_2 p'_1 p'_2) \mp t(p_1 p_2 p'_2 p'_1)|^2 \delta\left(E_{p_1} + E_{p_2} - E_{p'_1} - E_{p'_2}\right)\delta(p_1 + p_2 - p'_1 - p'_2), \tag{126}$$

which leads to the quantum statistical Boltzmann equation.

3.2.6 Properties of the Boltzmann equation

The Boltzmann equation is a nonlinear integro-differential equation for the single-particle distribution function in the classical case. In the quantum case, we can use the density matrix or the Wigner function to characterize the nonequilibrium state of the system. The Boltzmann equation is valid in low-density limit (only binary collisions). At higher densities also three-body collisions, etc., must be taken into account. Further density effects such as the formation of quasi particles and bound states have to be considered. The collision term is approximated to be local in space and time, no gradients in the density and no memory in time is considered. The assumption of molecular chaos means that correlations are neglected, the two-particle distribution function is replaced by the product of single-particle distribution functions.

The increase of entropy (Boltzmann H theorem) can be proven. In terms of the relevant statistical operator, the entropy is

$$S_{\rm rel} = k_{\rm B}\sum_p \{(\mp 1 + f_1(p))\ln(1 \mp f_1(p)) - f_1(p)\ln f_1(p)\}. \tag{127}$$

The change with time follows from

$$\begin{aligned}\frac{dS_{\rm rel}}{dt} &= -k_{\rm B}\sum_p \frac{\partial f_1}{\partial t}\ln f_1 - k_{\rm B}\sum_p \frac{\partial f_1}{\partial t} + k_{\rm B}\sum_p \frac{\partial f_1}{\partial t}\ln(1 \mp f_1) + k_{\rm B}\sum_p \frac{\partial f_1}{\partial t} \\ &= k_{\rm B}\sum_{p_1 p_2 p'_1 p'_2} w(p_1 p_2 p'_1 p'_2)\ln\left(\frac{1}{f_1(p_1)} \mp 1\right)\left\{\left(\frac{1}{f_1(p'_1)} \mp 1\right)\left(\frac{1}{f_1(p'_2)} \mp 1\right) - \left(\frac{1}{f_1(p_1)} \mp 1\right)\left(\frac{1}{f_1(p_2)} \mp 1\right)\right\} \\ &\times f_1(p_1) f_1(p'_1) f_1(p_2) f_1(p'_2).\end{aligned} \tag{128}$$

We interchange indices $1 \leftrightarrow 2, 1' \leftrightarrow 2'$; furthermore $1 \leftrightarrow 1', 2 \leftrightarrow 2'$; and $1 \leftrightarrow 2', 2 \leftrightarrow 1'$, use the symmetries of $w(\boldsymbol{p}_1\boldsymbol{p}_2\boldsymbol{p}'_1\boldsymbol{p}'_2)$ and $(x_1 - x_2)(\ln x_1 - \ln x_2) \geq 0$ because $\ln x$ is a monotonous function of x. We obtain $4\frac{\mathrm{d}S_{\mathrm{rel}}}{\mathrm{d}t} \geq 0$, the Boltzmann (relevant) entropy can increase.

The collision integral guarantees conservation of total momentum, particle number, and kinetic energy. However, the total energy including the interaction part is not conserved. The equilibrium solution $f_1^0(\boldsymbol{p})$ follows from $\frac{\mathrm{d}S_{\mathrm{rel}}}{\mathrm{d}t} = 0$:

$$\left(\frac{1}{f_1^0(\boldsymbol{p})} \mp 1\right)\left(\frac{1}{f_1^0(\boldsymbol{p}_1)} \mp 1\right) - \left(\frac{1}{f_1^0(\boldsymbol{p}')} \mp 1\right)\left(\frac{1}{f_1^0(\boldsymbol{p}'_1)} \mp 1\right) = 0. \tag{129}$$

If $f_1^0(\boldsymbol{p})$ depends only on energy, we find the well-known result for ideal quantum gases,

$$\frac{1}{f_1^0(\boldsymbol{p})} \mp 1 = \mathrm{e}^{\beta(E_p - \mu)}, \qquad f_1^0(\boldsymbol{p}) = \left[\mathrm{e}^{\beta(E_p - \mu)} \pm 1\right]^{-1}. \tag{130}$$

In the classical limit, we have $f_1^0(\boldsymbol{p}) = \mathrm{e}^{-\beta(E_p - \mu)}$ with $\mathrm{e}^{\beta\mu} = \frac{N}{\Omega}\left(\frac{2\pi\hbar^2}{mk_{\mathrm{B}}T}\right)^{3/2}\frac{1}{(2s+1)}$, where s denotes the spin of the particle.

3.2.7 Beyond the Boltzmann kinetic equation

In deriving the Boltzmann equation, different approximations have been performed: only binary collisions are considered, three-particle, and higher order collisions are neglected. Memory effects and spatial inhomogeneities have been neglected. The single-particle distribution was considered as relevant observable in the Markov approximation. These approximations can be compared with the Born-Markov approximation discussed in context with the quantum master equation. Instead of the Born approximation that is possible for weak interactions, the binary collision approximation is possible in the low-density limit, where three- and higher order collisions are improbable.

In the case of thermal equilibrium, the Boltzmann entropy S_{rel} (127) coincides with the entropy of the ideal (classical or quantum) gas. The equilibrium solution of the Boltzmann equation leads to the entropy of the ideal gas and gives not the correct equation of state for an interacting system that are derived from the Gibbs entropy ($\Phi = \ln Z$ is the Matthieu-Planck function)

$$S_{\mathrm{eq}} = -k_{\mathrm{B}} \int \mathrm{d}\Gamma(\Phi + \beta H) \exp\left[-\Phi - \beta H\right], \tag{131}$$

see Eq. (13). This deficit of the Boltzmann equation arises because binary collisions are considered where the kinetic energy of the asymptotic states is conserved. Only the single-particle distribution is a relevant observable and is correctly reproduced. It can be improved if the total energy, which is conserved, is considered as a relevant observable. Alternatively, we can also include the two-particle distribution function in the set of relevant observables. An important example is the formation of bound states as a signature of strong correlations in the system. Then, the momentum distribution of bound states has to be included in the set of relevant observables.

3.2.8 The linearized Boltzmann equation

Different approximations are known to obtain solutions of the Boltzmann equation, see [4, 5]. A serious problem in solving the Boltzmann equation is its nonlinearity as we have terms of the form $f_1(\boldsymbol{p}_1, t) f_1(\boldsymbol{p}_2, t)$. Special cases that allow for linearization are two-component systems with a large difference in the masses or concentration. Linearization is also possible in the case where the deviation from some equilibrium distribution is small. As an application, we consider the calculation of electrical conductivity in plasmas.

We investigate a plasma of ions and electrons under the influence of an external electric field $\boldsymbol{E}_{\text{ext}}$. For simplicity, we assume $\boldsymbol{E}_{\text{ext}}$ to be homogeneous and independent of time (statical conductivity σ). For moderate fields, we await a linear behavior of the plasma following Ohm's Law:

$$\boldsymbol{j}_{\text{el}} = \sigma \boldsymbol{E}. \tag{132}$$

[Note that in Eq. (132) $\boldsymbol{E}$ is not the external field, but the effective electric field in the medium (the plasma), being the superposition of the external field $\boldsymbol{E}_{\text{ext}}$ and the polarization field $\varepsilon \boldsymbol{P}$]. $\boldsymbol{j}_{\text{el}}$ is the average electric current defined via the single-particle distribution function f_1

$$\boldsymbol{j}_{el} = \frac{1}{\Omega} \left\langle \sum_i^N e_i \boldsymbol{v}_i \right\rangle = \sum_s e_s \int d^3 \boldsymbol{v} \boldsymbol{v} f_1(\boldsymbol{v}, s) = \sum_s \frac{e_s}{m_s} \int \frac{d^3 \boldsymbol{p}}{(2\pi\hbar)^3} \boldsymbol{p} f_1(\boldsymbol{p}, s). \tag{133}$$

Here, we have kept the index s for the different sorts. In the following, we will skip this index as we only consider electrons being responsible for the electric current.

We recall the Boltzmann equation

$$\frac{\boldsymbol{p}}{m} \frac{\partial}{\partial \boldsymbol{r}} f_1 + e\boldsymbol{E} \frac{\partial}{\partial \boldsymbol{p}} f_1 + \left(\frac{\partial}{\partial t} f_1 \right)_{\text{St}} = 0, \tag{134}$$

m is the electron mass and $-e$ the electron charge. The first term in this equation vanishes because of homogeneity of the system. For the collision term, we take the expression Eq. (125) in the generalized form for quantum systems. After the distribution function of the collision partner has been replaced by the equilibrium distribution, we have

$$\left(\frac{\partial}{\partial t} f_1 \right)_{\text{St}} = \int \frac{d^3 \boldsymbol{p}' \Omega}{(2\pi\hbar)^3} \{ f_1(\boldsymbol{p}') w_{\boldsymbol{p}\boldsymbol{p}'} [1 - f_1(\boldsymbol{p})] - f_1(\boldsymbol{p}) w_{\boldsymbol{p}'\boldsymbol{p}} [1 - f_1(\boldsymbol{p}')] \}, \tag{135}$$

where $w_{\boldsymbol{p}\boldsymbol{p}'}$ is the transition rate from the momentum state $\boldsymbol{p}$ to the state $\boldsymbol{p}'$. The quantum behavior of the collisions is taken into account via the Pauli blocking factors $[1 - f_1(\boldsymbol{p})]$.

3.2.9 Example: conductivity of the Lorentz plasma

In the Lorentz plasma model, the electron-electron collisions are neglected, and only electron-ion collisions are considered, interaction potential $V_{ei}(\boldsymbol{r})$. In the adiabatic approximation where the ions are regarded as fixed at positions $\boldsymbol{R}_i$ (elastic collisions), the interaction part of the Hamiltonian reads

$$\mathrm{H}' = \sum_i V_{ei}(\boldsymbol{r} - \boldsymbol{R}_i). \tag{136}$$

In Born approximation (or time-dependent perturbation theory), the transition rate is given by Fermi's Golden rule:

$$w_{p'p} = \frac{2\pi}{\hbar}\left|\mathrm{H}'_{p'p}\right|^2 \delta(E_p - E_{p'}) = w_{pp'}; \qquad E_p = p^2/2m. \tag{137}$$

To solve the Boltzmann equation Eq. (135), we make use of the ansatz

$$f_1(\boldsymbol{p}) = f_1^0(E_p) + \Phi(\boldsymbol{p})\frac{\mathrm{d}f_1^0(E_p)}{\mathrm{d}E_p}k_\mathrm{B}T = f_1^0(E_p)\{1 + \Phi(\boldsymbol{p})(1 - f_1^0(E_p))\}. \tag{138}$$

For equilibrium distributions, we have the detailed balance condition

$$w_{pp'}f_1^0(E_{p'})(1 - f_1^0(E_p)) = w_{p'p}f_1^0(E_p)(1 - f_1^0(E_{p'})). \tag{139}$$

Insertion of Eq. (138) into the Boltzmann equation Eq. (135) yields with Eq. (139)

$$\frac{e}{mk_\mathrm{B}T}\boldsymbol{E}\cdot\boldsymbol{p}f_1^0(E_p)\left[1 - f_1^0(E_p)\right] = \int\frac{\mathrm{d}^3\boldsymbol{p}'\Omega}{(2\pi\hbar)^3}w_{pp'}f_1^0(E_{p'})\left[1 - f_1^0(E_p)\right][\Phi(\boldsymbol{p}') - \Phi(\boldsymbol{p})], \tag{140}$$

where we have neglect terms with higher order of E and have used the fact that $\Phi(\mathbf{p}) \propto E$. With the definition of the *relaxation time tensor* $\hat{\tau}(\boldsymbol{p})$, according to $\Phi(\boldsymbol{p}) = e/(mk_\mathrm{B}T)\boldsymbol{E}\cdot\hat{\tau}(\boldsymbol{p})\cdot\boldsymbol{p}$, the equation reads

$$\boldsymbol{e}_\mathrm{E}\cdot\boldsymbol{p} = \int\frac{\mathrm{d}^3\boldsymbol{p}'\Omega}{(2\pi\hbar)^3}w_{pp'}\frac{f_1^0(E_{p'})}{f_1^0(E_p)}\boldsymbol{e}_\mathrm{E}\cdot(\hat{\tau}(\boldsymbol{p}')\cdot\boldsymbol{p}' - \hat{\tau}(\boldsymbol{p})\cdot\boldsymbol{p}), \tag{141}$$

$\boldsymbol{e}_\mathrm{E} = \boldsymbol{E}/E$. The electric current density Eq. (133) depends only on the deviation of the distribution function since f_1^0 is an even function in $\boldsymbol{p}$ (isotropy). We obtain by insertion of Eq. (138) into Eq. (133)

$$\boldsymbol{j}_\mathrm{el} = \frac{e}{\Omega}2\int\frac{\mathrm{d}^3\boldsymbol{p}\Omega}{(2\pi\hbar)^3}\frac{\boldsymbol{p}}{m}\Phi(\boldsymbol{p})f_1^0(E_p)\left[1 - f_1^0(E_p)\right]. \tag{142}$$

The conductivity σ is the proportionality factor between the current density and the effective field $\boldsymbol{E}$:

$$\sigma = \frac{e^2}{m^2k_\mathrm{B}T}2\int\frac{\mathrm{d}^3\boldsymbol{p}}{(2\pi\hbar)^3}p_z(\hat{\tau}(\boldsymbol{p})\cdot\boldsymbol{p})_z f_1^0(E_p)\left[1 - f_1^0(E_p)\right]. \tag{143}$$

We have derived an analytical expression for the conductivity of a Lorentz plasma in terms of the relaxation time tensor $\hat{\tau}(\boldsymbol{p})$. For isotropic systems, $\hat{\tau}_{ij} = \tau\delta_{ij}$, the well-known Ziman formula $\sigma_\tau = \tau ne^2/m$ for the conductivity results.

The solution of Eq. (141) for a momentum-dependent relaxation time is

$$\tau(E_p) = \left\{\int\frac{\mathrm{d}^3\boldsymbol{p}'\Omega}{(2\pi\hbar)^3}w_{pp'}(1 - \cos\vartheta)\right\}^{-1} \tag{144}$$

as can be verified by insertion. Now, the conductivity reads with Eq. (137)

$$\sigma = \frac{2e^2}{m^2 k_B T \Omega} \int d^3p\, p_z^2 f_1^0(E_p)\left[1 - f_1^0(E_p)\right] \left\{ \frac{2\pi}{\hbar} \int d^3p' |H_{p'p}|^2 \delta(E_p - E_{p'})(1 - \cos\vartheta) \right\}^{-1}. \quad (145)$$

Considering the screened interaction potential (Debye potential) $V_{ei}^{D}(r) = \frac{e^2}{4\pi\epsilon_0 |r|} e^{-\kappa|r|}$ with the Debye screening parameter $\kappa^2 = e^2 N/(\epsilon_0 k_B T \Omega)$, the evaluation can be performed. With

$$\Lambda(p) = \int_0^{2p/\hbar} \frac{1}{(q^2 + \kappa^2)^2} q^3 dq = \ln\sqrt{1+b} - \frac{1}{2}\frac{b}{1+b}, \qquad b = \frac{4p^2 k_B T \Omega \epsilon_0}{e^2 \hbar^2 N}, \quad (146)$$

we finally obtain for the conductivity [5].

$$\sigma = \frac{2^{5/2}(k_B T)^{3/2}(4\pi\epsilon_0)^2}{\pi^{3/2} m^{1/2} e^2 \Lambda}; \qquad \Lambda \approx \Lambda\left(p^2/2m = 3k_B T/2\right). \quad (147)$$

3.2.10 Conclusions

The method of the nonequilibrium statistical operator gives not only the derivation of the Boltzmann equation (quantal and classical), but indicates also possible improvements such as conservation of total energy, inclusion of bound state formation, hydrodynamic equations, etc.

The solution of the general Boltzmann equation is not simple, in addition to numerical simulations different approximations have been worked out. For the linearized Boltzmann equation, the relaxation time approximation can be used for elastic scattering, but for the general case (inclusion of electron-electron collisions in a plasma), the Kohler variational principle [11] can be applied. Landau-Vlasov equations for mean-field effects as well as Fokker-Planck equations for the collision term have been investigated.

The basic assumption to derive the Boltzmann equation is the selection of the single-particle distribution as relevant observable. Correlations are neglected and have to be built up in higher orders of approximation or extending the set of relevant observables. The most appropriate systems for kinetic theory are dilute gases where the collision time is short compared with the time of free flight. Irreversibility is owing to the Stoßzahlansatz for the intrinsic interaction.

3.3 Linear response theory

A third example, which allows the explicit elimination of the Lagrange multipliers to fulfill the self-consistency conditions, is a system near to thermodynamic equilibrium which is under the influence of mechanical (external forces) or thermodynamic (gradients of temperature, pressure, chemical potentials, etc.) perturbations. As response, currents appear in the system. Assuming linearity for small perturbations, transport coefficients are defined. Fluctuations in equilibrium are considered as a nonequilibrium state which relaxes to equilibrium, see Eq. (7).

3.3.1 Response to an external field

We consider a system under the influence of external (time dependent) fields acting on the particles, see [4, 11–16],

$$H^t = H_S + H_F^t, \quad (148)$$

where H_S denotes the system Hamiltonian, containing all kinetic energies of the particles as well as the full interaction part. The second part H_F^t describes the coupling of the system to the external fields h_j:

$$H_F^t = -\sum_j h_j e^{-i\omega t} A_j. \tag{149}$$

We characterize the nonequilibrium state by the set $\{B_n\}$ of relevant observables. In the following, we assume that the equilibrium expectation values of the nonequilibrium fluctuations disappear, $\langle B_n \rangle_{eq} = 0$ (else, we have to subtract the equilibrium values).

Treating the conserved observables explicitly, we write the relevant statistical operator ρ_{rel} in the form ($\mathcal{H} = H_S - \sum_c \mu_c N_c$)

$$\rho_{\text{rel}}(t) = e^{-\Phi(t) - \beta\left(\mathcal{H} - \sum_n F_n(t) B_n\right)}, \qquad \Phi(t) = \ln \mathrm{Tr}\left\{ e^{-\beta\left(\mathcal{H} - \sum_n F_n(t) B_n\right)} \right\}, \tag{150}$$

where the Lagrange multipliers are divided into the equilibrium parameters β, μ and the generalized response parameters $F_n(t)$, coupled to the corresponding observables. All Lagrange parameters are determined by the given mean values of these observables. In particular, we have the self-consistency conditions (18)

$$\langle B_n \rangle_{\text{rel}}^t = \mathrm{Tr}\{\rho_{\text{rel}}(t) B_n\} = \mathrm{Tr}\{\rho(t) B_n\} = \langle B_n \rangle^t \tag{151}$$

or

$$\mathrm{Tr}\{\rho_{\text{irrel}}(t) B_n\} = 0, \qquad \rho_{\text{irrel}}(t) = \rho(t) - \rho_{\text{rel}}(t). \tag{152}$$

The corresponding self-consistency condition for N and H_S lead to the well-known equations of state for the temperature $1/\beta$ and the chemical potential μ. $\Phi(t)$ is the Massieu-Planck functional that normalizes $\rho_{\text{rel}}(t)$.

We consider the limit of weak external fields. Compared with the equilibrium distribution (13), we expect that the changes of the state of the system are also weak. We characterize the nonequilibrium state by the set $\{B_n\}$ of relevant observables and assume that the averages

$$\langle B_n \rangle^t = \mathrm{Tr}\{\rho(t) B_n\} \propto h_j e^{-i\omega t} \tag{153}$$

are proportional to the external fields (linear response).

The basic assumption of LRT is that the average values $\langle B_n \rangle^t$ of the additional observables, which characterize the response of the system, are proportional to the external fields. Because these external fields are arbitrarily weak, we expand all quantities with respect to the fields up to first order. If the fluctuations $\langle B_n \rangle^t$ are proportional to these fields, we have also $F_n \propto h_j$. Below, we derive linear equations that relate the response of the system to the causing external fields.

In the linear regime, we await the response parameters $F_n(t)$ to exhibit the same time dependence as the external fields:

$$F_n(t) = F_n e^{-i\omega t}. \tag{154}$$

Here, we have harmonic fields $h_j e^{-i\omega t}$, but the formulation rests general as we can always express arbitrary time dependences by means of a Fourier transformation. Within the linear regime, the superposition of different components of the field gives the superposition of the corresponding responses. The treatment of spatial dependent external forces is also possible. As a specific advantage of the Zubarev method, thermodynamic forces such as gradients of temperature or chemical potentials can be treated [4, 5, 15, 16].

3.3.2 Elimination of the Lagrange multipliers

The main problem is to eliminate the Lagrange multipliers, the generalized response parameters $F_n(t)$. As in the case of kinetic theory, this is also possible explicitly in the case of linear response theory (LRT). With the operator relation $\mathrm{e}^{\mathrm{A}+\mathrm{B}} = \mathrm{e}^{\mathrm{A}} + \int_0^1 d\lambda\, \mathrm{e}^{\lambda(\mathrm{A}+\mathrm{B})}\mathrm{B}\,\mathrm{e}^{(1-\lambda)\mathrm{A}}$, we get for the relevant statistical operator (150) up to first order of the nonequilibrium fluctuations $\{\mathrm{B}_n\}$

$$\rho_{\mathrm{rel}}(t) = \rho_{\mathrm{eq}} + \beta \int_0^1 d\lambda \sum_n F_n(t)\, \mathrm{B}_n(\mathrm{i}\hbar\beta\lambda)\rho_{\mathrm{eq}}. \tag{155}$$

Here, we made use of the modified-Heisenberg picture $\mathrm{O}(\tau) = \exp(\mathrm{i}\mathcal{H}\tau/\hbar)\mathrm{O}\exp(-\mathrm{i}\mathcal{H}\tau/\hbar)$ with $\tau \to \mathrm{i}\hbar\beta\lambda$ replacing in the exponents $\mathrm{H_S}$ by $\mathcal{H} = \mathrm{H_S} - \sum_c \mu_c \mathrm{N}_c$. We want to calculate expectation values of macroscopic relevant variables that commute with the particle number operator N_c so that we can use both $\mathcal{H}$ and $\mathrm{H_S}$ synonymously. (Mention that also the Massieu-Planck functional $\Phi(t)$ has to be expanded so that the fluctuations around the equilibrium averages $\left\{\mathrm{B}_n - \langle \mathrm{B}_n\rangle_{\mathrm{eq}}\right\}$ appear).

3.3.3 Linearization of the NSO

All terms have to be evaluated in such a way, that the total expression rests of order $\mathcal{O}(h)$. For expressions (25) and (26), we find after integration by parts

$$\rho_\epsilon(t) = \rho_{\mathrm{rel}}(t) - \int_{-\infty}^{t} dt_1 \mathrm{e}^{\epsilon(t_1-t)}\mathrm{U}(t,t_1)\left\{\frac{\mathrm{i}}{\hbar}\left[\left(\mathrm{H_S}+\mathrm{H_F^{t_1}}\right),\rho_{\mathrm{rel}}(t_1)\right] + \frac{\partial}{\partial t_1}\rho_{\mathrm{rel}}(t_1)\right\}\mathrm{U}^\dagger(t,t_1). \tag{156}$$

Since $\mathrm{H_S}$ commutes with ρ_{eq} (equilibrium!), the curly bracket is of order $\mathcal{O}(h)$. In particular, we have for the first term the time derivative in the Heisenberg picture,

$$\frac{\mathrm{i}}{\hbar}\left[\mathrm{H_S}, \beta\int_0^1 d\lambda \sum_n F_n(t_1)\mathrm{B}_n(\mathrm{i}\lambda\beta\hbar)\rho_{\mathrm{eq}}\right] = \beta\int_0^1 d\lambda \sum_n F_n(t_i)\dot{\mathrm{B}}_n(i\lambda\beta\hbar)\rho_{\mathrm{eq}}. \tag{157}$$

For the second term of the integral in Eq. (156), we use Kubo's identity

$$\left[\mathrm{B}, \mathrm{e}^{\mathrm{A}}\right] = \int_0^1 d\lambda\, \mathrm{e}^{\lambda\mathrm{A}}[\mathrm{B},\mathrm{A}]\mathrm{e}^{(1-\lambda)\mathrm{A}}. \tag{158}$$

so that

$$\frac{\mathrm{i}}{\hbar}\left[\mathrm{H}_F^{t_1},\rho_{\mathrm{eq}}\right]=-\beta e^{-\mathrm{i}\omega t_1}\int_0^1 d\lambda\sum_j h_j\dot{\mathrm{A}}_j(\mathrm{i}\lambda\beta\hbar)\rho_{\mathrm{eq}}. \tag{159}$$

The last term in the curly bracket can be rewritten as

$$\frac{\partial}{\partial t_1}\rho_{\mathrm{rel}}=\beta\int_0^1 d\lambda\sum_n \dot{F}_n(t_1)\mathrm{B}_n(i\lambda\beta\hbar)\rho_{\mathrm{eq}}. \tag{160}$$

Because we restrict ourselves to the order $\mathcal{O}(h)$, for the time evolution operator we have $\mathrm{U}(t,t_1)\simeq \mathrm{e}^{-\mathrm{i}\mathrm{H}_{\mathrm{S}}(t-t_1)/\hbar}$.

After linearization with respect to the external fields h_j and the response parameters F_n, finally we have

$$\begin{aligned}\rho_\epsilon(t)=\rho_{\mathrm{rel}}(t)-\beta\mathrm{e}^{-\mathrm{i}\omega t}\int_{-\infty}^0 dt_1\mathrm{e}^{-\mathrm{i}z t_1}\int_0^1 d\lambda\Bigg[&-\sum_j h_j\dot{\mathrm{A}}_j(\mathrm{i}\lambda\beta\hbar+t_1)\rho_{\mathrm{eq}}\\ &+\sum_n\left(F_n\dot{\mathrm{B}}_n(\mathrm{i}\lambda\beta\hbar+t_1)\rho_{\mathrm{eq}}-\mathrm{i}\omega F_n\mathrm{B}_n(\mathrm{i}\lambda\beta\hbar+t_1)\rho_{\mathrm{eq}}\right)\Bigg]\end{aligned} \tag{161}$$

$(z=\omega+i\epsilon)$. Here, we used that $h_j(t)$ and $F_n(t)$, Eq. (154), are proportional to $\mathrm{e}^{-i\omega t}$.

We multiply this equation by B_m, take the trace and use the self-consistency relation (151). We obtain a set of linear equations for the thermodynamically conjugated parameters F_n (response parameters):

$$\sum_n\left\{\left\langle\mathrm{B}_m;\dot{\mathrm{B}}_n\right\rangle_z-\mathrm{i}\omega\langle\mathrm{B}_m;\mathrm{B}_n\rangle_z\right\}F_n=\sum_j\left\langle\mathrm{B}_m;\dot{\mathrm{A}}_j\right\rangle_z h_j, \tag{162}$$

with the Kubo scalar product (the particle number commutes with the observables)

$$(\mathrm{A}|\mathrm{B})=\int_0^1 d\lambda\mathrm{Tr}\left\{\mathrm{A}\mathrm{e}^{-\lambda\beta\mathcal{H}}\mathrm{B}\mathrm{e}^{\lambda\beta\mathcal{H}}\rho_{\mathrm{eq}}\right\}=\int_0^1 d\lambda\,\mathrm{Tr}\left\{\mathrm{A}\mathrm{B}(\mathrm{i}\lambda\beta\hbar)\rho_{\mathrm{eq}}\right\}, \tag{163}$$

and its Laplace transform, the thermodynamic correlation function

$$\langle\mathrm{A};\mathrm{B}\rangle_z=\int_{-\infty}^0 dt\,\mathrm{e}^{-\mathrm{i}zt}(\mathrm{A}|\mathrm{B}(t))=\int_0^\infty dt\,\mathrm{e}^{\mathrm{i}zt}(\mathrm{A}(t)|\mathrm{B}). \tag{164}$$

The linear system of equations (162) has the form

$$\sum_n P_{mn}F_n=\sum_j D_{mj}h_j \tag{165}$$

to determine the response parameters F_n, the number of equations coincides with the number of variables to be determined. The coefficients of this linear system of equations are given by equilibrium correlation functions. We emphasize that in the classical limit the relations become more simple because the variables commute, and we have not additional integrals expanding the exponential.

We can solve this linear system of equations (162) using Cramers rule. The response parameters F_n are found to be proportional to the external fields h_j with coefficients that are ratios of two determinants. The matrix elements are given by equilibrium correlation functions. This way, the self-consistency conditions are solved, and the Lagrange multipliers can be eliminated. The nonequilibrium problem is formally solved. The second problem, the evaluation of equilibrium correlation functions, can be solved by different methods such as numerical simulations, quantum statistical perturbation theories such as thermodynamic Green functions and Feynman diagrams, path integral methods, etc. Using partial integration, we show the relation

$$-\mathrm{i}z\langle \mathrm{A};\mathrm{B}\rangle_z = (\mathrm{A}|\mathrm{B}) + \langle \dot{\mathrm{A}};\mathrm{B}\rangle_z = (\mathrm{A}|\mathrm{B}) - \langle \mathrm{A};\dot{\mathrm{B}}\rangle_z. \tag{166}$$

Then, the generalized linear response equations (162) can be rewritten in the short form (165) with the matrix elements

$$P_{mn} = \left(\mathrm{B}_m|\dot{\mathrm{B}}_n\right) + \langle \dot{\mathrm{B}}_m;\dot{\mathrm{B}}_n\rangle_{\omega+\mathrm{i}\epsilon} - \mathrm{i}\omega(\mathrm{B}_m|\mathrm{B}_n) - \mathrm{i}\omega\langle \dot{\mathrm{B}}_m;\mathrm{B}_n\rangle_{\omega+\mathrm{i}\epsilon}, \tag{167}$$

$$D_{mj} = \left(\mathrm{B}_m|\dot{\mathrm{A}}_j\right) + \langle \dot{\mathrm{B}}_m;\dot{\mathrm{A}}_j\rangle_{\omega+\mathrm{i}\epsilon}. \tag{168}$$

that can be interpreted as generalized transition rates (collision integral, left-hand side) and the influence of external forces (drift term, right-hand side of Eq. (165)).

Having the response parameters F_n to our disposal, we can evaluate averages of the relevant observables, see Eq. (151),

$$\langle \mathrm{B}_n\rangle^t = \langle \mathrm{B}_n\rangle^t_{\mathrm{rel}} = -\beta\sum_m F_m \mathrm{e}^{\mathrm{i}\omega t} N_{mn}, \qquad N_{mn} = (\mathrm{B}_m|\mathrm{B}_n). \tag{169}$$

Eliminating F_m, these average fluctuations $\langle \mathrm{B}_n\rangle^t$ are proportional to the fields $h_j \mathrm{e}^{-\mathrm{i}\omega t}$.

3.3.4 Force-force correlation function and static (dc) conductivity

As an example for the generalized linear response theory, we calculate the conductivity of a plasma of charged particles (electrons and ions) that is exposed to a static homogeneous electric field in x-direction: $\omega = 0$, $\boldsymbol{E} = E\boldsymbol{e}_x$,

$$\mathrm{H_F} = -eE\mathrm{X}, \qquad \mathrm{X} = \sum_i^{N_e} \mathrm{x}_i. \tag{170}$$

Instead of h_j, we have only one constant external field E. For the treatment of arbitrary ω to obtain the dynamical (optical) conductivity see [11, 13, 16, 17]. The conjugated variable A from Eq. (149) that couples the system to the external field is $\mathrm{A} = e\mathrm{X}$. The time derivative follows as $\dot{\mathrm{A}} = (e/m)\mathrm{P}$, with $\mathrm{P} = \sum_i^{N_e} \mathrm{p}_{x,i}$ denoting the total momentum in x direction.

For simplicity, the ions are considered here as fixed in space because of the large mass ratio (adiabatic approximation). Then, the transport of charge is owing to the motion of the electrons. In general, the ions can also be treated as moving charged particles that contribute to the current.

A stationary state will be established in the plasma where the electrons are accelerated by the external field, but loose energy (and momentum) due to collisions with the ions. This nonequilibrium state is characterized by an electrical current that is absent in thermal equilibrium. We can take the electric current density $\mathrm{j}_{el} = (e/m\Omega)\mathrm{P} = (e/\Omega)\dot{\mathrm{X}}$ as a relevant observable that characterizes the nonequilibrium state. Instead, we take the total momentum $\mathrm{B} = \mathrm{P} = m\dot{\mathrm{X}}$. The generalized linear response equations (165) and (167) read

$$F\left[(\dot{\mathrm{P}}|\mathrm{P}) + \langle \dot{\mathrm{P}};\dot{\mathrm{P}}\rangle_{ic}\right] = \frac{e}{m}E\left\{(\mathrm{P}|\mathrm{P}) + \langle \mathrm{P};\dot{\mathrm{P}}\rangle_{i\epsilon}\right\}, \tag{171}$$

The term $(\dot{\mathrm{P}}|\mathrm{P}) = \langle[\mathrm{P},\mathrm{P}]\rangle_{eq}$ vanishes as can be shown with Kubo's identity, see Eq. (158). With the Kubo identity, we also evaluate the Kubo scalar product

$$(\mathrm{P}|\mathrm{P}) = m\int_0^1 \mathrm{d}\lambda\langle\dot{\mathrm{X}}(-\mathrm{i}\hbar\beta\lambda)\mathrm{P}\rangle_{\mathrm{eq}} = -\frac{im}{\hbar\beta}\mathrm{Tr}\left\{\rho_{\mathrm{eq}}[\mathrm{X},\mathrm{P}]\right\} = \frac{mN}{\beta}. \tag{172}$$

The solution for response parameter F is

$$F = \frac{e}{m}E\frac{mN/\beta + \langle \mathrm{P};\dot{\mathrm{P}}\rangle_{i\epsilon}}{\langle \dot{\mathrm{P}};\dot{\mathrm{P}}\rangle_{i\epsilon}}. \tag{173}$$

With Eq. (169) we have

$$\langle \mathrm{j}_{\mathrm{el}}\rangle = \frac{e}{m\Omega}\langle \mathrm{P}\rangle_{\mathrm{rel}} = \frac{e\beta}{m\Omega}F(\mathrm{P}|\mathrm{P}) = \sigma_{\mathrm{dc}}E. \tag{174}$$

The resistance R in the static limit follows as

$$R = \frac{1}{\sigma_{\mathrm{dc}}} = \frac{\Omega\beta}{e^2N^2}\frac{\langle \dot{\mathrm{P}};\dot{\mathrm{P}}\rangle_{i\epsilon}}{1 + \langle \mathrm{P};\dot{\mathrm{P}}\rangle_{i\epsilon}\beta/mN}. \tag{175}$$

3.3.5 Ziman formula for the Lorentz plasma

To evaluate the resistance R, we have to calculate the correlation functions $\langle \dot{\mathrm{P}};\dot{\mathrm{P}}\rangle_{i\epsilon}$ and $\langle \mathrm{P};\dot{\mathrm{P}}\rangle_{i\epsilon}$. For this, we have to specify the system Hamiltonian $\mathrm{H_S}$, which reads for the Lorentz plasma model (136)

$$\mathrm{H_S} = \mathrm{H_0} + \mathrm{H_{int}} = \sum_p E_p \mathrm{a}_p^\dagger \mathrm{a}_p + \sum_{p,q} V_q \mathrm{a}_{p+q}^\dagger \mathrm{a}_p, \qquad E_p = \frac{\hbar^2p^2}{2m}. \tag{176}$$

We consider the ions at fixed positions $\boldsymbol{R}_i$ so that $V(\boldsymbol{r}) = \sum_i V_{\mathrm{ei}}(\boldsymbol{r} - \boldsymbol{R}_i)$. The Fourier transform V_q depends for isotropic systems only on the modulus $q = |\boldsymbol{q}|$ and will be specified below. A realistic plasma Hamiltonian should consider also moving ions and the electron-electron interaction so that we have a two-component plasma Hamiltonian with pure Coulomb interaction between all constituents. This has been worked out [14], but is not the subject of our present work so that we restrict ourselves mainly to the simple Lorentz model.

The force $\dot{\mathrm{P}}$ on the electrons follows from the x component of the total momentum ($\boldsymbol{p}$ is the wave-number vector)

$$\mathrm{P} = \sum_p \hbar p_x \mathrm{a}_p^\dagger \mathrm{a}_p, \qquad [\mathrm{H_S}, \mathrm{P}] = -\sum_{p,q} V_q \hbar q_x \mathrm{a}_{p+q}^\dagger \mathrm{a}_p. \tag{177}$$

We calculate the force-force correlation function (only x component)

$$\langle \dot{\mathrm{P}}; \dot{\mathrm{P}} \rangle_{i\epsilon} = \int_{-\infty}^{0} \mathrm{d}t\, \mathrm{e}^{\epsilon t} \int_0^1 \mathrm{d}\lambda \left\langle \frac{\mathrm{i}}{\hbar} [\mathrm{H_S}, \mathrm{P}(t - \mathrm{i}\lambda\beta\hbar)] \frac{\mathrm{i}}{\hbar} [\mathrm{H_S}, \mathrm{P}] \right\rangle_{\mathrm{eq}} \tag{178}$$

in Born approximation with respect to V_q. In lowest order, the force-force correlation function is of second order so that in the time evolution $\exp[(\mathrm{i}/\hbar)\mathrm{H_S}(t - \mathrm{i}\lambda\beta\hbar)]$ the contribution $\mathrm{H_{int}}$ of interaction to $\mathrm{H_S}$, Eq. (176), can be dropped as well as in the statistical operator. The averages are performed with the noninteracting ρ_0. The product of the two commutators is evaluated using Wick's theorem. One obtains

$$\begin{aligned} \langle \dot{\mathrm{P}}; \dot{\mathrm{P}} \rangle_{i\epsilon} &= -\sum_{p,p',q,q'} \int_{-\infty}^{0} \mathrm{d}t\, \mathrm{e}^{\epsilon t} \int_0^1 \mathrm{d}\lambda\, \mathrm{e}^{\frac{\mathrm{i}}{\hbar}(E_p - E_{p+q})(t - \mathrm{i}\hbar\beta\lambda)} V_q V_{q'} q_x q'_x \left\langle \mathrm{a}_{p+q}^\dagger \mathrm{a}_p \mathrm{a}_{p'+q'}^\dagger \mathrm{a}_{p'} \right\rangle_{\mathrm{eq}} \\ &= \sum_{p,q} |V_q|^2 \delta(E_p - E_{p+q}) f_p \left(1 - f_p\right) \pi\hbar q_x^2. \end{aligned} \tag{179}$$

Because the x direction can be arbitrarily chosen in an isotropic system, we replace $q_x^2 = \left(q_x^2 + q_y^2 + q_z^2\right)/3 = q^2/3$ if the remaining contributions to the integrand are not depending on the direction in space.

Evaluating Eq. (175) in Born approximation, the correlation function $\langle \mathrm{P}; \dot{\mathrm{P}} \rangle_{i\epsilon} (\beta/mN)$ can be neglected in relation to 1 because it contains the interaction strength. For the resistance, this term contributes only to higher orders of the interaction.

The force-force correlation function (179) is further evaluated using the relations $-\frac{1}{\beta}\frac{\mathrm{d}f(E_p)}{\mathrm{d}E_p} = f_p\left(1 - f_p\right)$ and $\delta(E_p - E_{p+q}) = \frac{m}{\hbar^2 qp}\delta\left(\cos\theta - \frac{q}{2p}\right)$. The q integration has to be performed in the limits $0 \le q \le 2p$. Finally the resistance can be calculated by inserting the previous expressions Eqs. (172) and (179) into Eq. (175) so that the Ziman-Faber formula is obtained,

$$R = \frac{m^2 \Omega^3}{12\pi^3 \hbar^3 e^2 N^2} \int_0^\infty \mathrm{d}E(p) \left(-\frac{\mathrm{d}f(E)}{\mathrm{d}E}\right) \int_0^{2p} \mathrm{d}q\, q^3 |V_q|^2. \tag{180}$$

The expression for the resistance depends on the special form of the potential V_q. For a pure Coulomb potential $e^2/(\Omega\epsilon_0 q^2)$, the integral diverges logarithmically as typical for Coulomb integrals. The divergency at very small values of q is removed if screening due to the plasma is taken into account. Within a many-particle approach, in static approximation the Coulomb potential is replaced by the Debye potential (146). The evaluation yields

$$\sigma_{\mathrm{dc}} = \frac{3}{4\sqrt{2\pi}} \frac{(k_B)^{3/2} (4\pi\epsilon_0)^2}{m^{1/2} e^2} \frac{1}{\Lambda(p_{\mathrm{therm}})} \tag{181}$$

where the Coulomb logarithm is approximated by the value of the average p, with $\hbar^2 p_{\mathrm{therm}}^2/2m = 3k_B T/2$. In the low-density limit, the asymptotic behavior of the Coulomb logarithm Λ is given by $-(1/2)\ln n$. However, this result for σ_{dc} is not correct and can only be considered as an approximation, as discussed below considering the virial expansion of the resistivity.

3.3.6 Different sets of relevant observables

After fully linearizing the statistical operator (161) with (155), we have for the electrical current density

$$\langle j_{el} \rangle = \frac{e}{m\Omega} \langle P \rangle = \frac{e\beta}{m\Omega} \left\{ \sum_n \left[(P|B_n) - \langle P; \dot{B}_n \rangle_{i\epsilon} \right] F_n + \langle P; P \rangle_{i\epsilon} \frac{e}{m} E \right\} = \sigma_{dc} E. \quad (182)$$

After deriving the Ziman formula from the force-force correlation function in the previous section, we investigate the question to select an appropriate set of relevant observables $\{B_n\}$.

3.3.7 Kubo formula

The most simple choice of relevant observables is the empty set. There are no response parameters to be eliminated. According Eq. (182), the Kubo formula

$$\sigma_{dc}^{Kubo} = \frac{e^2 \beta}{m^2 \Omega} \langle P; P \rangle_{i\epsilon}^{irred} \quad (183)$$

follows [18, 19]. The index "irred" denotes the irreducible part of the correlation function, because the conductivity is not describing the relation between the current and the external field, but the internal field. We will not discuss this in the present work. A similar expression can also be given for the dynamical, wave-number vector-dependent conductivity $\sigma(\boldsymbol{q}, \omega)$ which is related to other quantities such as the response function, the dielectric function, or the polarization function, see [5, 11, 16, 17]. Equation (183) is a fluctuation-dissipation theorem; equilibrium fluctuations of the current density are related to a dissipative property, the electrical conductivity.

The idea to relate the conductivity with the current-current auto-correlation function in thermal equilibrium looks very appealing because the statistical operator is known. The numerical evaluation by simulations can be performed for any densities and degeneracy. However, the Kubo formula (183) is not appropriate for perturbation theory. In the lowest order of interaction, we have the result $\sigma_{dc}^{Kubo,\,0} = ne^2/m\epsilon$ (conservation of total momentum) which diverges in the limit $\epsilon \to 0$.

3.3.8 Force-force correlation function

The electrical current can be considered as a relevant variable to characterize the nonequilibrium state, when a charged particle system is affected by an electrical field. We can select the total momentum as the relevant observable, $B_n \to P$. Now, the character of Eq. (182) is changed. According the response equation (162), we have

$$-\langle P; \dot{P} \rangle_{i\epsilon} F + \langle P; P \rangle_{i\epsilon} \frac{e}{m} E = 0 \quad (184)$$

so that these contributions compensate each other. As a relevant variable, the averaged current density is determined by the response parameter F which follows from the solution of the response equation (184). We obtain the inverse conductivity, the resistance, as a force-force auto-correlation, see Eq. (175). Now, perturbation theory can be applied, and in Born approximation a standard result of transport theory is obtained, the Ziman formula (180). We conclude that the use of relevant observables gives a better starting point for perturbation theory. In contrast to the Kubo formula that starts from thermal equilibrium as initial state, the correct current is already reproduced in the initial state and must not be created by the dynamical evolution.

However, despite the excellent results using the Ziman formula in solid and liquid metals where the electrons are strongly degenerated, we cannot conclude that the result (181) for the conductivity is already correct for low-density plasmas (nondegenerate limit if T remains constant) in the lowest order of perturbation theory considered here. The prefactor $3/(4\sqrt{2\pi})$ is wrong. If we go to the next order of interaction, divergent contributions arise. These divergences can be avoided performing a partial summation, that will also change the coefficients in Eq. (181) which are obtained in the lowest order of the perturbation expansion. The divergent contributions can also be avoided extending the set of relevant observables $\{B_n\}$, see below.

3.3.9 Higher moments of the single-particle distribution function

Besides the electrical current, also other deviations from thermal equilibrium can occur in the stationary nonequilibrium state, such as a thermal current. In general, for homogeneous systems, we can consider a finite set of moments of the single-particle distribution function

$$P_n = \sum_p \hbar p_x (\beta E_p)^{n/2} a_p^\dagger a_p \tag{185}$$

as set of relevant observables $\{B_n\}$. It can be shown that with increasing number of moments the result

$$\sigma_{\rm dc} = s \frac{(k_B)^{3/2}(4\pi\epsilon_0))^2}{m^{1/2}e^2} \frac{1}{\Lambda(p_{\rm therm})} \tag{186}$$

is improved, as can be shown with the Kohler variational principle, see [11, 15]. The value $s = 3/(4\sqrt{2\pi})$ obtained from the single moment approach is increasing to the limiting value $s = 2^{5/2}/\pi^{3/2}$. For details see [5, 15, 16], where also other thermoelectric effects in plasmas are considered.

3.3.10 Single-particle distribution function and the general form of the linearized Boltzmann equation

Kinetic equations are obtained if the occupation numbers n_ν of single-(quasi-) particle states $|\nu\rangle$ is taken as the set of relevant observables $\{B_n\}$. The single-particle state ν is described by a complete set of quantum numbers, e.g., the momentum, the spin and the species in the case of a homogeneous multi-component plasma. In thermal equilibrium, the averaged occupation numbers of the quasiparticle states are given by the Fermi or Bose distribution function, $\langle n_\nu\rangle_{\rm eq} = f_\nu^0 = {\rm Tr}\{\rho_{\rm eq} n_\nu\}$. These equilibrium occupation numbers are changed under the influence of the external field. We consider the deviation $\Delta n_\nu = n_\nu - f_\nu^0$ as relevant observables. They describe the fluctuations of the occupation numbers. The response equations, which eliminate the corresponding response parameters F_ν, have the structure of a linear system of coupled Boltzmann equations for the quasiparticles, see [11]

$$\frac{e}{m}\boldsymbol{E}\cdot\left[(\mathbf{P}|\mathrm{n}_\nu)+\langle\mathbf{P};\dot{\mathrm{n}}_\nu\rangle_{i\epsilon}\right]=\sum_{\nu'}F_{\nu'}P_{\nu'\nu}, \tag{187}$$

with $P_{\nu'\nu}=(\dot{\mathrm{n}}_{\nu'}|\Delta\mathrm{n}_\nu)+\langle\dot{\mathrm{n}}_{\nu'};\dot{\mathrm{n}}_\nu\rangle_{i\epsilon}$. The response parameters F_ν are related to the averaged occupation numbers as

$$f_\nu(t)=\mathrm{Tr}\{\rho(t)\mathrm{n}_\nu\}=f_\nu^0+\beta\sum_{\nu'}F_{\nu'}(\Delta\mathrm{n}_{\nu'}|\Delta\mathrm{n}_\nu). \tag{188}$$

The general form of the linear Boltzmann equation (187) can be compared with the expression obtained from kinetic theory. The left-hand side can be interpreted as the drift term, where self-energy effects are included in the correlation function $\langle\mathbf{P};\dot{\mathrm{n}}_\nu\rangle_{i\epsilon}$. Because the operators n_ν are commuting, from the Kubo identity follows $(\dot{\mathrm{n}}_{\nu'}|\mathrm{n}_\nu)=(1/\hbar\beta)\langle[\mathrm{n}_{\nu'},\mathrm{n}_\nu]\rangle=0$. In the general form, the collision operator is expressed in terms of equilibrium correlation functions of fluctuations that can be evaluated by different many-body techniques. In particular, for the Lorentz model the result (186) with $s=2^{5/2}/\pi^{3/2}$ is obtained [5, 15, 16].

3.3.11 Two-particle distribution function, bound states

Even more information is included if we also consider the nonequilibrium two-particle distributions. As an example, we mention the Debye-Onsager relaxation effect, see [5, 14]. Another important case is the formation of bound states. It seems naturally to consider the bound states as new species and to include the occupation numbers (more precisely, the density matrix) of the bound particle states in the set of relevant observables [20, 21]. It needs a long memory time to produce bound states from free states dynamically in a low-density system, because bound states cannot be formed in binary collisions, a third particle is needed to fulfill the conservation laws.

The inclusion of initial correlation to improve the kinetic theory, in particular to fulfill the conservation of total energy, is an important step worked out during the last decades, see [22] where further references are given. Other approaches to include correlations in the kinetic theory are given, e.g., in [23, 24].

3.3.12 Conclusions

Transport coefficients are expressed in terms of correlation functions in equilibrium. The evaluation can be performed numerically (molecular-dynamic simulations), or using quantum statistical methods such as perturbation theory and the technique of Green functions. The generalized linear response theory has solved problems owing to the evaluation of correlation functions. Perturbation expansions are improved if higher orders are considered. The treatment of singular terms that appear in perturbation expansions is quite complex. Alternatively, the set of relevant observables can be extended. Examples are the virial expansion of the conductivity [14] or the hopping conductivity [5, 12].

It is not clear whether the rigorous evaluation of the correlation functions (i.e., the limit $\epsilon\to 0$ only after full summation of the perturbation expansion) will give nontrivial results for the conductivity. For instance, arguments can be given that the exact evaluation of the force-force correlation function to calculate the resistance leads to a vanishing result, and the correlation function of stochastic forces must be considered, in analogy to the corresponding term in the Langevin equation [6, 25]. A related projection operator technique was used by Mori [26] for the memory-function approach.

There are close relation to other approaches, such as kinetic theory or quantum master equations, where the response function of the bath is considered. Irreversibility is not inherent in the equilibrium correlation functions, but in the assumption that a nonequilibrium state is considered as a fluctuation in equilibrium with a prescribed value of the relevant quantity. Other degrees of freedom are forced to adopt the distribution of thermal equilibrium.

4. Concluding remarks

4.1 Information theory

The method of nonequilibrium statistical operator (NSO) to describe irreversible processes is based on a very general concept of entropy, the Shannon information entropy (10). This concept is not restricted to dynamical properties like energy, particle numbers, momentum, etc., occurring in physics, but may be applied also to other properties occurring, e.g., in economics, financial market, and game theory. The generalized Gibbs distributions (13) and (19) are obtained if the averages of a given set of observables are known. Other statistical ensembles may be constructed, where the values of some observables have a given distribution. For instance, the canonical ensemble follows if the particle numbers are fixed, and the microcanonical ensemble has in addition a fixed energy in the interval ΔE around E, see [1, 2]. There exist alternative concepts of entropy to valuate a probability distribution which are not discussed here.

In physics, we have a dynamical evolution that forms the equilibrium distribution for ergodic systems, and any initial distribution that is compatible with the values of the conserved quantities can be used to produce the correct equilibrium distribution. The main problem is the microscopic approach to evaluate the dynamical averages, which can be done using quantum statistical methods such as Green function theory or path integral calculations, or, alternatively, numerical simulations of the microscopic equations of motion such as molecular dynamics. In more general, complex systems, we do not know the exact dynamics of the time evolution. However, we can observe time-dependent correlation functions which reflect the time evolution, and properties such as the fluctuation-dissipation theorem are not related to a specific dynamical model for the complex system. The most interesting issue of the NSO method is the selection of the set of relevant observables to describe a nonequilibrium process. The better the choice of the set of relevant observables is, for which a dynamical model for the time evolution can be found, the less influence is produced by the irrelevant observables which may be described by time-dependent correlation functions.

4.2 Hydrodynamics

An important application is the description of hydrodynamic processes and its relation to kinetic theory. The NSO method allows to treat this problem, selecting the single-particle distribution as well as the hydrodynamic variables as set of relevant observables. This approach has been worked out in [23]. A more general presentation is found in [4], and transport processes in multi-component fluids and superfluid systems are investigated. Until now, a rigorous theory of turbulence is not available, but hydrodynamic fluctuations and turbulent flow have been considered using the NSO method [4].

4.3 The limit $\epsilon \to 0$

It is the source term of the extended von Neumann equation (27) that introduces irreversible behavior. Different choices for the set of relevant observables are elected for different applications, in particular quantum master equations, kinetic theory, and linear response theory. It is claimed that this choice of the set of relevant observables is only a technical issue and has no influence on the result, only if the limit $\varepsilon \to 0$ is correctly performed in the final result.

However, calculations are not performed this way. For instance, the limit $\varepsilon \to 0$ is performed already in a finite order of perturbation theory. The self-consistency conditions (18) guarantee that a finite source term will not influence the Hamiltonian dynamics of the relevant observables. A closer investigation of a finite source term and its influence on the nonequilibrium evolution would be of interest.

4.4 Heat production and entropy

A serious problem is that irreversibility is connected with the production of entropy [6]. For instance, in the case of electrical conductivity, heat is produced. In principle, we have to consider an open system coupled to a bath that absorbs the produced heat. In the Zubarev NSO method considered here, it is the right-hand of the extended von Neumann equation (27) that contains the source term. We impose the stationary conditions so that ρ_{rel}, in particular T, are not explicitly depending on time. Then, the source term acts like an additional process describing the coupling to a bath without specifying the microscopic process. The parameter ϵ now has the meaning of a relaxation time, and is no longer arbitrarily small but is of the order E^2.

From a systematic microscopic point of view, one can introduce a process into the system Hamiltonian which describes the cooling of the system via the coupling to a bath, as known from the quantum master equations for open systems. Phonons related to the motion of ions can be absorbed by the bath, but one can calculate the electrical conductivity also for (infinitely) heavy ions so that the scattering of the electrons, accelerated by the field, is elastic. Collisions of electrons with the bath may help, but an interesting process to reduce the energy is radiation. Electrons which are accelerated during the collisions emit bremsstrahlung. This heat transfers the gain of energy of electrons, which are moving in the external field, to the surroundings.

4.5 Open systems: coupling to the radiation field

A general approach to scattering theory was given by Gell-Mann and Goldberger [27] (see also [1, 2]) to incorporate the boundary condition into the Schrödinger equation. The equation of motion in the potential $\mathrm{V}(\boldsymbol{r})$ reads

$$\frac{\partial}{\partial t}\psi_\epsilon(\boldsymbol{r},t)+\frac{\mathrm{i}}{\hbar}\mathrm{H}\psi_\epsilon(\boldsymbol{r},t)=-\epsilon\left[\psi_\epsilon(\boldsymbol{r},t)-\psi^{\hat{t}}_{\text{rel}}(\boldsymbol{r},t)\right]. \tag{189}$$

With $\mathrm{H}=\mathrm{H}_0+\mathrm{V}$, the relevant state is an eigenstate $|\boldsymbol{p}\rangle$ of H_0 which changes its value at the scattering time $\hat{t}$ where the asymptotic state $|\boldsymbol{p}'\rangle$ is formed. As known from the Langevin equation, one can consider $\psi_\epsilon(\boldsymbol{r},t)=\varrho^{1/2}\exp(\mathrm{i}S/\hbar)$ as a stochastic process [5] related to a stochastic potential $\mathrm{V}(\boldsymbol{r},t)$; Eq. (189) appears as an average. The relaxation term is related to the fluctuations of $\mathrm{V}(\boldsymbol{r},t)$. The average

Hamiltonian dynamics is realized by the self-consistency conditions for $\psi_{\text{rel}}^{\hat{t}}(\boldsymbol{r},t)$, see Eq. (28).

An interesting example is the electrical conductivity. In the stationary case which is homogeneous in time, the system remains near thermodynamic equilibrium as long as the electrical field is weak so that the produced heat can be exported. We have to consider an open system. If the conductor is embedded in vacuum, heat export is given by radiation. Bremsstrahlung is emitted during the collision of charged particles. Emission of photons can be considered as a measuring process to localize the charged particle during the collision process. The time evolution of the system is considered as a stochastic process, see also [6].

Author details

Gerd Röpke
Institute of Physics, University of Rostock, Rostock, Germany

*Address all correspondence to: gerd.roepke@uni-rostock.de

References

[1] Zubarev DN. Nonequilibrium Statistical Thermodynamics. New York: Plenum Press; 1974

[2] Zubarev DN. The statistical operator for nonequilibrium systems. Doklady Akademii Nauk SSSR. 1961;**140**:92

[3] Zubarev D, Morozov V, Röpke G. Statistical Mechanics of Nonequilibrium Processes. Vol. 1. Berlin: Akademie-Verlag; 1996

[4] Zubarev D, Morozov V, Röpke G. Statistical Mechanics of Nonequilibrium Processes. Vol. 2. Berlin: Akademie-Verlag; 1997

[5] Röpke G. Nonequilibrium Statistical Physics. Weinheim: Wiley-VCH; 2013

[6] Röpke G. Electrical conductivity of charged particle systems and zubarevs nonequilibrium statistical operator method. Theoretical and Mathematical Physics. 2018;**194**:74

[7] Boltzmann L. Vorlesungen über Gastheorie. Vol. 2. Leipzig: Verlag; 1898

[8] Bogoliubov NN. Problems of Dynamic Theory in Statistical Physics (in Russian). Moscow-Leningrad: Gostekhizdat; 1946

[9] Gocke C, Röpke G. Master equation of the reduced statistical operator of an atom in a plasma. Theoretical and Mathematical Physics. 2008;**154**:26

[10] Lin C, Gocke C, Röpke G, Reinholz H. Transition rates for a Rydberg atom surrounded by a plasma. Physical Review A. 2016;**93**:042711

[11] Reinholz H, Röpke G. Dielectric function beyond the random-phase approximation: Kinetic theory versus linear response theory. Physical Review E. 2012;**85**:036401

[12] Christoph V, Röpke G. Theory of inverse linear response coefficients. Physica Status Solidi B. 1985;**131**:11

[13] Reinholz H, Redmer R, Röpke G, Wierling A. Long-wavelength limit of the dynamical local-field factor and dynamical conductivity of a two-component plasma. Physical Review E. 2000;**62**:5648

[14] Röpke G. Quantum-statistical approach to the electrical conductivity of dense, high-temperature plasmas. Physical Review A. 1988;**38**:3001

[15] Redmer R. Physical properties of dense, low-temperature plasmas. Physics Reports. 1997;**282**:35

[16] Reinholz H. Dielectric and optical properties of dense plasmas. Annales de Physique (Paris). 2005;**30**:1

[17] Röpke G. Dielectric function and electrical dc conductivity of nonideal plasmas. Physical Review E. 1998;**57**:4673

[18] Kubo R. Statistical-Mechanical theory of irreversible processes. Journal of the Physical Society of Japan. 1957;**12**: 570

[19] Kubo R. The fluctuation-dissipation theorem. Reports on Progress in Physics. 1966;**29**:255

[20] Röpke G. Electrical conductivity of a system of localized and delocalized electrons. Theoretical and Mathematical Physics. 1981;**46**:184

[21] Adams J et al. Coulomb contribution to the direct current electrical conductivity of dense partially ionized plasmas. Physics of Plasmas. 2007;**14**: 062303

[22] Morozov VD et al. The "Mixed" green's function approach to quantum

kinetics with initial correlations. Annals of Physics (New York). 1999;**278**:127

[23] Zubarev DN, Morozov VG, Omelyan IP, Tokarchuk MV. Unification of the kinetic and hydrodynamic approaches in the theory of dense gases and liquids. Theoretical and Mathematical Physics. 1993;**96**:997

[24] Morozov VG, Röpke G. Non-Markovian quantum kinetics and conservation laws. Journal of Statistical Physics. 2001;**102**:285

[25] Kalashnikov VP. Linear relaxation equations in the nonequilibrium statistical operator method. Theoretical and Mathematical Physics. 1978;**34**:263

[26] Mori H. A continued-fraction representation of the time-correlation functions. Progress in Theoretical Physics. 1965;**34**:399

[27] Gell-Mann M, Goldberger ML. The formal theory of scattering. Physics Review. 1953;**91**:398

Chapter 2

The Boundary Element Method for Fluctuating Active Colloids

William E. Uspal

Abstract

The boundary element method (BEM) is a computational method particularly suited to solution of linear partial differential equations (PDEs), including the Laplace and Stokes equations, in complex geometries. The PDEs are formulated as boundary integral equations over bounding surfaces, which can be discretized for numerical solution. This manuscript reviews application of the BEM for simulation of the dynamics of "active" colloids that can self-propel through liquid solution. We introduce basic concepts and model equations for both catalytically active colloids and the "squirmer" model of a ciliated biological microswimmer. We review the foundations of the BEM for both the Laplace and Stokes equations, including the application to confined geometries, and the extension of the method to include thermal fluctuations of the colloid. Finally, we discuss recent and potential applications to research problems concerning active colloids. The aim of this review is to facilitate development and adoption of boundary element models that capture the interplay of deterministic and stochastic effects in the dynamics of active colloids.

Keywords: active colloids, Brownian dynamics, boundary element method

1. Introduction

Over the past 15 years, significant effort has been invested in the development of synthetic micro- and nano-sized colloids capable of self-propulsion in liquid solution [1–3]. These "active colloids" have myriad potential applications in drug delivery [4, 5], sensing [6], microsurgery [7], and programmable materials assembly [8]. Furthermore, they provide well-controlled model systems for study of materials systems maintained out of thermal equilibrium by continuous dissipation of free energy. In this context, and in comparison with driven systems (e.g., sheared suspensions), a unique aspect of active colloids is that energy is injected into the system at the *microscopic* scale of a single particle, instead of through macroscopic external fields or at the boundaries of the system. As a consequence of this, novel collective behaviors are possible, including motility-induced phase separation [9], mesoscopic "active turbulence" [10], and formation of dynamic "living crystals" and clusters [11, 12]. Furthermore, since living systems can be regarded as self-organized non-equilibrium materials systems, study of active colloids could yield insight into fundamental principles of living systems, and open a path towards development of biomimetic "dissipative materials" capable of homeostasis [13], self-repair [14], goal-directed behavior [15, 16], and other aspects of life.

IntechOpen

Paradigmatic examples of synthetic active colloids include bimetallic Janus rods [17] and Janus spheres consisting of a spherical core with a hemispherical coating of a catalytic material [18]. In both cases, self-propulsion is driven by catalytic decomposition of a chemical "fuel" available in the liquid solution. For instance, for gold/platinum Janus rods, both ends of the rod are involved in the electrochemical decomposition of hydrogen peroxide into water and oxygen: hydrogen peroxide is oxidized at the platinum anode and reduced at the gold cathode. In this reaction process, a hydrogen ion gradient is established between the anode and cathode. The resulting gradient in electrical charge creates an electric field in the vicinity of the rod. The electric field exerts a force on the diffuse layer of ions surrounding the colloid surface, resulting in motion of the suspending fluid relative to the colloid surface. Viewed in a stationary reference frame, the final result is "self-electrophoretic" motion of the colloid in direction of the platinum end. For Janus spheres (e.g., platinum on silica or platinum on polystyrene), the mechanism of motion is still a subject of debate. Since the core material is inert and insulating, it was originally thought that these particles move by neutral self-diffusiophoresis in a self-generated oxygen gradient. Diffusiophoresis is similar to electrophoresis in that motion is driven by interfacial molecular forces. Briefly, in diffusiophoresis, the colloid surface and solute molecules interact through some molecular potential. This interaction potential, in conjunction with a gradient of solute concentration along the surface of the colloid, leads to the pressure gradient in a thin film surrounding the colloid, and therefore fluid flow within the film relative to the colloid surface. Following initial studies on chemically active Janus spheres, subsequent studies revealed a dependence of the Janus particle speed on the concentration of added salt [19], suggesting that a self-electrophoretic mechanism may be implicated in motion of the colloid. Golestanian and co-workers proposed that dependence of the rate of catalysis on thickness of the deposited catalyst can lead to different regions of the catalyst acting as anode and cathode [20]. More recently, it was proposed that if one of the redox reactions is reaction-limited and the other is diffusion-limited, the anodic or cathodic character of a point on the catalytic surface will depend on the local curvature of the surface [21]. Regardless of the detailed molecular mechanism of motion, a key point is that *interfacial flows* drive self-propulsion of chemically active colloids. A second key point is that particles need to have an intrinsic asymmetry (e.g., from the Janus character of their material composition) in order to exhibit directed motion.

These findings have motivated development of theoretical and numerical concepts for modeling the interfacially driven self-propulsion of active colloids. Motivated by classical work on phoresis in thermodynamic gradients [22, 23], an influential continuum framework for modeling neutral self-diffusiophoresis was established in Ref. 24, and will be reviewed below. This basic framework can be modified or extended to account for electrochemical effects [25], multicomponent diffusion [26], reactions in the bulk solution [27], and confinement [28–34]. An emerging area of study within this framework is autonomous navigation and "taxis" of chemically active colloids in ambient fields and complex geometries, including chemotaxis in chemical gradients [35] and rheotaxis in confined flows [15, 36]. Theoretical research on synthetic active colloids has also found common ground with an older strand of research on locomotion of *biological* microswimmers. Here, an important point of contact is again the idea of interfacial flow [37]. For a quasi-spherical microswimmer that is "carpeted" with a layer of cilia, the effect of the periodic, time-dependent, metachronal motion of the cilia can be modeled as a period-averaged interfacial flow. This "squirmer" model of locomotion was introduced by Lighthill [38] and refined by Blake [39]. More recent work has explored collective motion of suspensions of squirmers [40] and squirmer motion in confined geometries [41].

These theoretical frameworks are deterministic, and do not directly address the role of thermal fluctuations. For instance, for the model of a chemically active colloid in Ref. 42, diffusion of the chemical reaction product (i.e., the solute) into the surrounding solution is modeled with the Laplace equation, which has a smooth and unique solution for a given set of boundary conditions describing surface catalysis. Implicit in the use of the Laplace equation are the assumptions that, on the timescale of Janus particle motion, the solute diffuses very fast, and that fluctuations of the solute distribution average out to be negligible. Likewise, fluctuations of the surrounding fluid are neglected, i.e., the deterministic Stokes equation is used to model the fluid in lieu of the fluctuating Stokes equation. On the other hand, micron-sized active Janus particles are observed in experiments to exhibit "enhanced diffusion": directed motion on short timescales $t < \tau_r$ and random walk behavior on long timescales $t \gg \tau_r$. For the latter, the effective diffusion coefficient D_{eff} is enhanced relative to the "bare" diffusion coefficient D_0 of an inactive colloid, i.e., $D_{eff} \gg D_0$. The reason for this behavior is that the orientation of the particle is free to fluctuate, and the particle changes its direction of motion by rotational diffusion over the timescale $\tau_r = D_r^{-1}$, where D_r is the rotational diffusion coefficient of the particle [18]. Therefore, thermal fluctuations qualitatively affect the motion of even a micron-sized catalytic Janus particle in unbounded, uniform solution. For a catalytic Janus particle in an ambient field or in confinement, thermal fluctuations affect whether and for how long the particle can align with the ambient field [42, 43] or stay near confining surfaces [34, 44]. Overall, a full theoretical understanding of the behavior of micron-sized active colloids requires modeling thermal fluctuations.

Moreover, as part of the general drive towards miniaturization, recent experimental efforts have sought to fabricate and characterize nano-sized chemically active colloids [45–47]. On the theoretical side, new questions arise when the size of the colloid becomes comparable to the size of the various molecules participating in the catalytic reaction. These questions include: When is using a continuum model appropriate [48]? Can a catalytic particle still display (time- and ensemble-averaged) directed motion when the particle and the surrounding chemical field are fluctuating on similar timescales? Relatedly, can a spherical colloid with a catalytic surface of *uniform* composition exhibit enhanced diffusion when nano-sized [49]? Can a fluctuating, nano-sized Janus particle effectively follow an ambient chemical gradient, i.e., exhibit chemotaxis [35]? These questions also connect with the burgeoning literature on chemotaxis of biological enzymes [50].

In this chapter, we review the boundary element approach to modeling the motion of active colloids. This is a "hydrodynamic" approach that resolves the detailed geometry and surface chemistry of the colloids, the velocity of the surrounding solution, and the distribution of chemical species within the solution [30, 40, 51–57]. The advantage of such an approach—in comparison with, for instance, the active Brownian particle model—is that it can resolve the detailed microscopic physics of how a colloid couples to ambient fields and other features of the surrounding micro-environment. In addition, we discuss how thermal fluctuations can be included within the approach. The aim of this review is to facilitate development and adoption of models that capture the interplay of deterministic and stochastic effects within an integrated framework.

2. Theory

As a starting point, we review the basic deterministic theoretical framework for understanding the motion of active colloids [24]. This is a continuum approach that coarse-grains the interfacial flow that drives colloid motion, discussed above, as a "slip velocity" boundary condition for the velocity of the suspending fluid.

We consider a suspension of N active colloids in an unbounded liquid solution. The position of each colloid α, with $\alpha \in \{1, 2, ..., N\}$, is described in a stationary reference frame by a vector $\mathbf{x}_\alpha$. The solution is modeled as an incompressible Newtonian fluid with dynamic viscosity μ. The solution is governed by the Stokes equation,

$$-\nabla P + \mu \nabla^2 \mathbf{u} = 0, \tag{1}$$

where $P(\mathbf{x})$ is the pressure at a position $\mathbf{x}$ in the solution, and $\mathbf{u}(\mathbf{x})$ is the velocity of the solution. The velocity obeys the incompressibility condition,

$$\nabla \cdot \mathbf{u} = 0, \tag{2}$$

and the boundary condition

$$\mathbf{u}(\mathbf{x}_s) = \mathbf{U}_\alpha + \boldsymbol{\Omega}_\alpha \times (\mathbf{x}_s - \mathbf{x}_\alpha) + \mathbf{v}_{s,\alpha}(\mathbf{x}_s), \quad \mathbf{x}_s \in \mathcal{S}_\alpha, \tag{3}$$

where $\mathcal{S}_\alpha$ is the surface of colloid α, $\mathbf{x}_s \in \mathcal{S}_\alpha$ is a position on $\mathcal{S}_\alpha$, and $\mathbf{U}_\alpha$ and $\boldsymbol{\Omega}_\alpha$ are the translational and rotational velocities, respectively, of colloid α. The quantity $\mathbf{v}_{s,\alpha}(\mathbf{x}_s)$ is the slip velocity on the surface of colloid α, which is either prescribed (for a squirmer) or determined by the distribution of chemical species in solution (for a chemically active colloid). The form of $\mathbf{v}_{s,\alpha}(\mathbf{x}_s)$ for the two types of particles will be discussed in detail below. Additionally, far away from the N particles, the fluid velocity vanishes:

$$\mathbf{u}(|\mathbf{x}| \to \infty) = 0. \tag{4}$$

In order to close this system of equations, we require $6N$ more equations, corresponding to the $6N$ unknown components of $\mathbf{U}_\alpha$ and $\boldsymbol{\Omega}_\alpha$. The net force and torque on each colloid vanishes:

$$\int_{\mathcal{S}_\alpha} \boldsymbol{\sigma} \cdot \hat{\mathbf{n}} \; dS + \mathbf{F}_{\text{ext},\alpha} = 0, \tag{5}$$

$$\int_{\mathcal{S}_\alpha} (\mathbf{x}_s - \mathbf{x}_\alpha) \times \boldsymbol{\sigma} \cdot \hat{\mathbf{n}} \; dS + \mathbf{T}_{\text{ext},\alpha} = 0, \tag{6}$$

where the integrals are performed over the surface $\mathcal{S}_\alpha$ of each colloid α, and $\mathbf{F}_{\text{ext},\alpha}$ and $\mathbf{T}_{\text{ext},\alpha}$ are, respectively, the net external force and net external torque on the colloid. The stress tensor is given by

$$\boldsymbol{\sigma} = -P\mathbf{I} + \mu\left(\nabla\mathbf{u} + (\nabla\mathbf{u})^T\right), \tag{7}$$

where the pressure $P(\mathbf{x})$ is determined by the incompressibility condition.

Practitioners of Stokesian Dynamics may notice some similarity between Eq. 3 and the boundary condition for an inert or passive sphere in an ambient flow field. If $\mathbf{v}_{s,\alpha}(\mathbf{x}_s)$ could be expressed as an effective ambient flow field at the position of particle α, the tools of Stokesian dynamics could be straightforwardly applied to simulation of active suspensions. This analogy will be developed in the Appendix.

2.1 The squirmer model: prescribed surface slip

The "squirmer" model was originally introduced by Lighthill to describe the time-averaged motion of ciliated quasi-spherical micro-organisms [38]. Lighthill's formulation was subsequently corrected and extended by Blake [39]. The basic motivating idea of the squirmer model is that the periodic, metachronal motion of the carpet of cilia on the surface of the micro-organism drives, over the course of one period and in the vicinity of the microswimmer surface, net flow from the "forward" or "leading" pole of the micro-organism to the "rear" pole (see **Figure 1**, left). This interfacial flow drives flow in the surrounding bulk fluid, leading to directed motion of the micro-organism towards the forward end. The squirmer model captures some essential features of the self-propulsion of micro-organisms, including the hydrodynamic interactions between micro-organisms, and between an individual micro-organism and confining surfaces.

The slip velocity on the surface of a spherical squirmer α is specified by fiat and does not depend on the configuration of the suspension. It is given as [41]:

$$\mathbf{v}_{s,\alpha}(\mathbf{x}_s) = \sum_n B_{n,\alpha} V_n(\cos\theta)\,\hat{\mathbf{e}}_{\theta_{p,\alpha}}, \tag{8}$$

where

$$V_n(x) \equiv \frac{2\sqrt{1-x^2}}{n(n+1)} \frac{dP_n(x)}{dx}. \tag{9}$$

The unit vector $\hat{\mathbf{e}}_{\theta_{p,\alpha}}$ is defined in following manner (see **Figure** 1, left). The squirmer has an axis of symmetry, and a propulsion direction $\hat{\mathbf{d}}_\alpha$ oriented along this axis. We define the body-frame polar angle $\theta_{p,\alpha}$ at a point $\mathbf{x}_s$ on the surface of squirmer α as the angle between $\hat{\mathbf{d}}_\alpha$ and a vector $\mathbf{x}_s - \mathbf{x}_\alpha$ from the center of the squirmer $\mathbf{x}_\alpha$ to $\mathbf{x}_s$. The unit vector $\hat{\mathbf{e}}_{\theta_{p,\alpha}}$ is oriented in the direction of increasing $\theta_{p,\alpha}$, i.e., locally tangent to the squirmer surface along a longitudinal line.

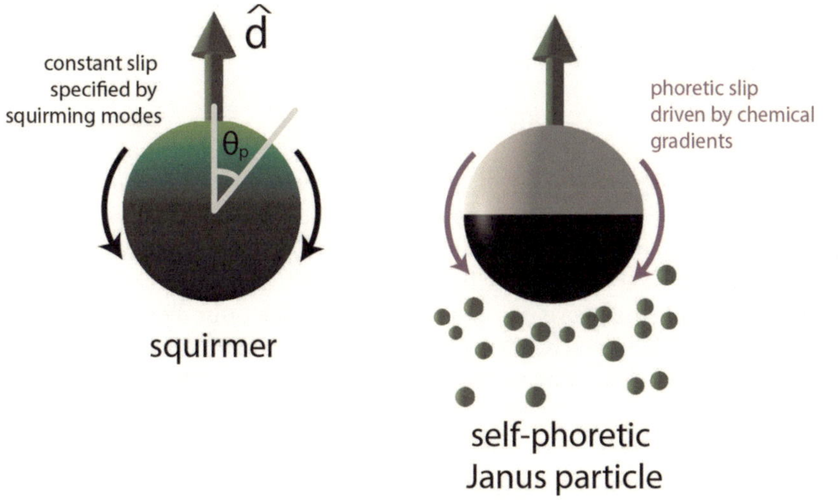

Figure 1.
The two types of microswimmer considered in this chapter. In the spherical "squirmer" model (left), the slip velocity on the surface of the particle is specified by fiat and fixed (in a frame co-moving and co-rotating with the particle) for all time. For an axisymmetric distribution of surface slip, the particle moves in the direction of the green arrow, i.e., opposite to the (surface-averaged) direction of the slip velocity. For a Janus particle (right), a fraction of the particle surface (black) catalyzes a reaction involving various molecular species diffusing in the surrounding solution. The resulting anisotropic distribution of product molecules or "solute" (green spheres) drives a phoretic slip velocity (purple) in an interfacial layer surrounding the particle. For a repulsive interaction between the solute and the particle surface, the slip is towards high concentration of solute, and the particle moves in the direction of the green arrow.

The squirming mode amplitudes $B_{n,\alpha}$, which can potentially vary from squirmer to squirmer, are fixed *a priori* and do not depend on the configuration of the suspension. The set of amplitudes determine the detailed form of the flow field in vicinity of the particle. Furthermore, the lowest order squirming mode B_1 determines the velocity of an isolated squirmer in unbounded solution: $U_{fs,\alpha} = (2/3)B_{1,\alpha}$. According to our definition of $\mathbf{d}_\alpha$ and $\theta_{p,\alpha}$, we require that $B_{1,\alpha}>0$. Simulations of squirmers typically truncate Eq. 8 to $n \leq 2$ or $n \leq 3$. The justification for this is that the contributions of the higher order squirming modes to the flow around the squirmer decay rapidly with distance from the squirmer.

2.2 Chemically active colloids: diffusiophoretic slip from chemical gradients

For chemically active colloids, the slip velocity on the surface of a colloid is driven by interfacial molecular forces. The molecular physics of phoresis and self-phoresis is reviewed in detail elsewhere [2, 23, 58]; here, we provide a brief summary. Consider a "Janus" colloid with a surface composed of two different materials. In the presence of molecular "fuel" diffusing in the surrounding solution, one of the two Janus particle materials catalyzes the decomposition of the fuel into molecular reaction products. A paradigmatic example of this reaction is the decomposition of hydrogen peroxide by platinum into water and oxygen:

$$H_2O_2 \underset{Pt}{\rightarrow} H_2O + \frac{1}{2}O_2. \tag{10}$$

(This equation is a severe simplification of the actual reaction scheme, which most likely involves charged and complex intermediates [20, 27]; nevertheless, proceeding from it, we can capture some essential features of self-phoresis.) If the reaction is reaction-limited—i.e., hydrogen peroxide is plentifully available in solution, and diffuses quickly relatively to the reaction rate—then we can approximate the production of oxygen with zero order kinetics:

$$-D[\nabla c \cdot \hat{\mathbf{n}}]|_{\mathbf{x}=\mathbf{x}_s} = \kappa(\mathbf{x}_s), \tag{11}$$

where D is the diffusion coefficient of oxygen, $c(\mathbf{x})$ is the number density of oxygen, and $\kappa(\mathbf{x}_s)$ is the rate of oxygen production on the surface of the particle. (The validity of assumption of reaction-limited kinetics is quantified by the Damköhler number $Da = \kappa_0 R/D$, where κ_0 is a characteristic reaction rate; we assume $Da \ll 1$.) Furthermore, we assume that the Péclet number $Pe \equiv U_0 R/D$ is very small, where U_0 is a characteristic particle velocity and R is the particle radius. Accordingly, we can make a quasi-steady approximation for the diffusion of oxygen in the solution:

$$\nabla^2 c = 0. \tag{12}$$

Finally, we assume that

$$c(|\mathbf{x}| \to \infty|) = c_\infty, \tag{13}$$

where c_∞ is a constant. Eqs. 11, 12, and 13 specify a boundary value problem (BVP) for the distribution of oxygen in the fluid domain containing the N active particles. This problem can be solved numerically, e.g., by the boundary element method, as will be outlined in a later section.

Accordingly, each Janus particle will be surrounded by an anisotropic "cloud" of oxygen molecules ("solute"), with the oxygen concentration highest near the catalytic cap (see **Figure 1**, right). Now we suppose that the oxygen molecules interact with the surface of the colloid through some molecular interaction potential with range $\delta \ll R$ [23]. Each colloid is surrounded by an interfacial layer of thickness $\sim \delta$ in which the molecular interaction of the solute and the colloid is significant. Outside of this layer, the solute freely diffuses in the solution. We can regard $c(\mathbf{x})$ as the bulk concentration, i.e., the concentration outside the interfacial layer. Near a location $\mathbf{x}_s$ on the surface of the colloid, the interfacial layer concentration is enhanced or depleted, according to the attractive or repulsive character of the molecular interaction, relative to $c(\mathbf{x}_s^+)$. Here, the plus sign emphasizes that $c(\mathbf{x}_s^+)$ is evaluated outside the interfacial layer. Moreover, since $\delta \ll R$, the interfacial layer concentration can *locally*, in the direction locally normal to the colloid surface, relax to a Boltzmann (i.e., equilibrium) distribution governed by the molecular interaction potential Φ. (The timescale for this local relaxation is much faster than the characteristic timescale for colloid motion R/U_0.) Accordingly, the local pressure $P(\mathbf{x}_s^-, \eta)$ can be calculated from Φ and $c(\mathbf{x}_s^+)$, where η is a coordinate defined at x_s that is locally normal to the colloid surface.

These notions can be made mathematically rigorous through the theory of matched asymptotics. However, for the purpose of this discussion, the essential idea is that the bulk concentration $c(\mathbf{x})$ determines the pressure in the interfacial layer in the vicinity of a point $\mathbf{x}_s$ on the colloid surface. Moreover, $c(\mathbf{x})$ varies over the length scale R of the colloid. Accordingly, within the interfacial layer, the pressure varies over the size of the colloid, driving flow within the interfacial layer. From the perspective of the outer solution for the flow field, this interfacial flow looks like a slip velocity:

$$\mathbf{v}_{s,\alpha}(\mathbf{x}_s) = -b(\mathbf{x}_s)\nabla_{\|}c. \tag{14}$$

Here, the surface gradient operator is defined as $\nabla_{\|} \equiv (\mathbf{I} - \hat{\mathbf{n}}\hat{\mathbf{n}}) \cdot \nabla$. The material-dependent parameter $b(\mathbf{x}_s)$ encapsulates the details of the molecular interaction between the solute and the surface material, and can be calculated from the molecular potential Φ [23]. Since the surface of the Janus colloid comprises different materials, b depends on the location on the colloid surface. In fact, a spatial variation of b over the surface of colloid is a necessary condition to obtain phoretic rotation of a colloid near a wall [30] or chemotactic alignment with a gradient of "fuel" molecules [35].

2.3 Lorentz reciprocal theorem

The Lorentz reciprocal theorem relates the fluid stresses $(\boldsymbol{\sigma}, \boldsymbol{\sigma}')$ and velocity fields $(\mathbf{u}, \mathbf{u}')$ of two solutions to the Stokes equation within the same domain $\mathcal{V}$:

$$\int_S \mathbf{u} \cdot \boldsymbol{\sigma}' \cdot \hat{\mathbf{n}}\ dS = \int_S \mathbf{u}' \cdot \boldsymbol{\sigma} \cdot \hat{\mathbf{n}}\ dS, \tag{15}$$

where $\mathcal{S}$ is the boundary of $\mathcal{V}$. For the N active particles in unbounded solution, $\mathcal{S} = \cup_{\alpha=1}^{N} \mathcal{S}_\alpha$.

This theorem can be used to simplify the problem specified above for the velocities of N active particles. We designate that problem as the "unprimed" problem. Additionally, we specify that $\mathbf{F}_{\text{ext},\alpha} = 0$ and $\tau_{\text{ext},\alpha} = 0$ for all α. (Since the Stokes equation is linear, the contributions of the external forces and torques to the velocities of the particles can be calculated separately, using standard methods, and superposed with the contributions from activity to obtain the complete velocities.) We consider $6N$ "primed" problems, indexed by $j = 1, 2, ..., 6N$. For problem $j = \alpha$, particle α is exposed to an external force with magnitude F'_{ext} in the $\hat{x}$ direction. Each particle has unknown translational and rotational velocities $\mathbf{U}'^{(j)}_{\alpha}$ and $\mathbf{\Omega}'^{(j)}_{\alpha}$, and the velocity field $\mathbf{u}'^{(j)}$ is subject to no-slip boundary conditions on each particle:

$$\mathbf{u}'^{(j)}(\mathbf{x_s}) = \mathbf{U}'^{(j)}_{\alpha} + \mathbf{\Omega}'^{(j)}_{\alpha} \times (\mathbf{x}_s - \mathbf{x}_\alpha), \quad \mathbf{x}_s \in \mathcal{S}_\alpha, \tag{16}$$

Additionally, the flow field vanishes far away from the particles, i.e., $\mathbf{u}'^{(j)}(|\mathbf{x}| \to \infty) = 0$. The unprimed problem and primed problem α are schematically illustrated in **Figure 2**. Similarly, for problems $j = \alpha + 1$ and $j = \alpha + 2$, particle α is exposed to an external force with magnitude in the $\hat{y}$ and $\hat{z}$ directions, respectively, with no-slip boundary conditions likewise holding on each particle, and the flow field vanishing far away from the particles. For problems $j = \alpha + 3, j = \alpha + 4$, and $j = \alpha + 5$, particle α is exposed to a torque with magnitude τ'_{ext} in the $\hat{x}, \hat{y}$, and $\hat{z}$ directions, respectively, with the same boundary conditions. Each "primed" problem j can be solved for $6N$ unknown velocity components $\mathbf{U}'^{(j)}_{\alpha}$ and $\mathbf{\Omega}'^{(j)}_{\alpha}$.

For problem j, we substitute Eqs. 3 and 16 into Eq. 15 for $\mathbf{u}$ and $\mathbf{u}' = \mathbf{u}'^{(j)}$ to obtain:

$$\sum_\alpha \int_{\mathcal{S}_\alpha} [\mathbf{U}_\alpha + \mathbf{\Omega}_\alpha \times (\mathbf{x}_s - \mathbf{x}_\alpha) + \mathbf{v}_{s,\alpha}(\mathbf{x}_s)] \cdot \boldsymbol{\sigma}'^{(j)} \cdot \hat{\mathbf{n}} \; dS = \tag{17}$$

$$\sum_\alpha \int_{\mathcal{S}_\alpha} [\mathbf{U}'_\alpha(j) + \mathbf{\Omega}'_\alpha(j) \times (\mathbf{x}_s - \mathbf{x}_\alpha)] \cdot \boldsymbol{\sigma} \cdot \hat{\mathbf{n}} \; dS. \tag{18}$$

It can be shown that the right hand side of this equation vanishes. Consider the term involving $\mathbf{U}'_\alpha(j)$. For each integral over $\mathcal{S}_\alpha$, $\mathbf{U}'_\alpha(j)$ is a constant and can be moved out of the integral,

$$\mathbf{U}'_\alpha(j) \cdot \int_{\mathcal{S}_\alpha} \boldsymbol{\sigma} \cdot \hat{\mathbf{n}} \; dS, \tag{19}$$

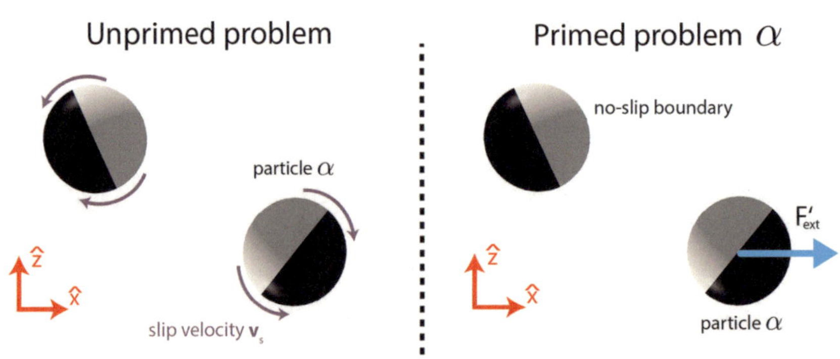

Figure 2.
Illustration of the "unprimed" problem for the velocities of N active particles, and the "primed" problem α for the velocities of N inert particles when particle α is exposed to a force with magnitude F'_{ext} in the $\hat{x}$ direction. Similarly, in primed problems $\alpha + 1$ and $\alpha + 2$, particle α is exposed to a force with magnitude F'_{ext} in the $\hat{y}$ direction and the $\hat{z}$ direction, respectively. Moreover, in primed problems $\alpha + 3$, $\alpha + 4$, and $\alpha + 5$, particle α is exposed to torques with magnitude τ'_{ext} in the $\hat{x}$, $\hat{y}$, and $\hat{z}$ directions, respectively. Note that the primed and the 6N unprimed problems all have the same geometry, i.e., the same configuration of N spheres.

but the integral is simply the force $\mathbf{F}_\alpha$ on particle α. Since the particles are free of external forces, $\mathbf{F}_\alpha = 0$. Likewise, the term involving $\mathbf{\Omega}'_\alpha(j)$ can be rearranged as

$$\mathbf{\Omega}'_\alpha(j) \cdot \int_{S_\alpha} (\mathbf{x}_s - \mathbf{x}_\alpha) \times \boldsymbol{\sigma} \cdot \hat{\mathbf{n}} \; dS, \tag{20}$$

but the integral is the torque $\boldsymbol{\tau}_\alpha = 0$ on particle α.

Rearranging the left hand side of Eq. 17, we obtain a set of 6N equations j:

$$\sum_\alpha \left(\mathbf{U}_\alpha \cdot \mathbf{F}'^{(j)}_\alpha + \mathbf{\Omega}_\alpha \cdot \boldsymbol{\tau}'^{(j)}_\alpha \right) = -\sum_\alpha \int_{S_\alpha} \mathbf{v}_s \cdot \boldsymbol{\sigma}'^{(j)} \cdot \hat{\mathbf{n}} \; dS \;, j = 1, \ldots, 6N \;. \tag{21}$$

These 6N equations can be written in matrix form:

$$\mathcal{R} \cdot \mathbf{V} = \mathbf{b}, \tag{22}$$

where $\mathbf{b}$ is a vector containing the 6N integrals associated with the right hand side of Eq. 21, $\mathbf{V}$ is vector of all 6N velocity components $(\mathbf{U}_\alpha, \mathbf{\Omega}_\alpha)^T$, and $\mathcal{R}$ is the grand resistance matrix for N spheres at positions $\mathbf{x}_\alpha$. Note that the arbitrary magnitudes F'_{ext} and τ'_{ext} have been divided out of Eq. 22.

The advantage of the reciprocal theorem approach is that if we solve the "primed" problem for a given set of particle positions $\mathbf{x}_\alpha$, we can easily compute the set $\mathbf{V}$ of 6N velocities for any set of slip velocities $\mathbf{v}_{s,\alpha}$. This, for instance, facilitates studying how various choices for $b(\mathbf{x}_s)$ or $B_{n,\alpha}$ affect particle motion. Additionally, the "primed" problem for the resistance matrix $\mathcal{R}$ and stresses $\boldsymbol{\sigma}'^{(j)}$ in a system of N spheres is a standard problem in microhydrodynamics. An interesting open question is whether this approach is numerically more stable than *directly* solving for the 6N particle velocities in the presence of the force-free and torque-free constraints.

2.3.1 Proof of Lorentz reciprocal theorem

We provide a short proof of Eq. 15, following the lines of Ref. 59 because some intermediate results will be useful later in the chapter. We recall that the rate of strain tensor e_{ij} is defined as

$$e_{ij} = \frac{1}{2}\left(\frac{\partial u_i}{\partial x_j} + \frac{\partial u_j}{\partial x_i} \right), \tag{23}$$

and that, in index notation, the stress tensor is

$$\sigma_{ij} = -P\delta_{ij} + 2\mu e_{ij} \tag{24}$$

We consider the quantity $\sigma'_{ij} e_{ij}$:

$$\begin{aligned} \sigma'_{ij} e_{ij} &= \left(-P'\delta_{ij} + 2\mu e_{ij}{}'\right) e_{ij} \\ &= -P' e_{ii} + 2\mu e'_{ij} e_{ij} \\ &= 2\mu e'_{ij} e_{ij}, \end{aligned} \tag{25}$$

where we have used $\nabla \cdot u = 0$ to eliminate e_{ii}, and we assume the Einstein convention for summation over repeated indices. Similarly, we can obtain $\sigma_{ij} e'_{ij} = 2\mu e'_{ij} e_{ij}$, so that

$$\sigma'_{ij}e_{ij} = \sigma_{ij}e'_{ij}. \tag{26}$$

We can also manipulate $\sigma'_{ij}e_{ij}$ as follows:

$$\sigma'_{ij}e_{ij} = \frac{1}{2}\sigma'_{ij}\frac{\partial u_i}{\partial x_j} + \frac{1}{2}\sigma'_{ij}\frac{\partial u_j}{\partial x_i}. \tag{27}$$

Swapping the two indices in the last term,

$$\sigma'_{ij}e_{ij} = \frac{1}{2}\sigma'_{ij}\frac{\partial u_i}{\partial x_j} + \frac{1}{2}\sigma'_{ji}\frac{\partial u_i}{\partial x_j}. \tag{28}$$

But $\sigma'_{ij} = \sigma'_{ji}$, giving

$$\begin{aligned} \sigma'_{ij}e_{ij} &= \sigma'_{ij}\frac{\partial u_i}{\partial x_j} \\ &= \frac{\partial}{\partial x_j}\left(\sigma'_{ij}u_i\right) - \left(\frac{\partial \sigma'_{ij}}{\partial x_j}\right), \end{aligned} \tag{29}$$

so that

$$\frac{\partial}{\partial x_j}\left(\sigma'_{ij}u_i\right) - \left(\frac{\partial \sigma'_{ij}}{\partial x_j}\right)u_i = \frac{\partial}{\partial x_j}\left(\sigma_{ij}u'_i\right) - \left(\frac{\partial \sigma_{ij}}{\partial x_j}\right)u'_i. \tag{30}$$

If there are no point forces applied to the fluid in determination of $\mathbf{u}'$ and $\mathbf{u}$, then $\nabla \cdot \boldsymbol{\sigma} = 0$ and $\nabla \cdot \boldsymbol{\sigma}' = 0$, and we obtain

$$\nabla \cdot (\mathbf{u} \cdot \boldsymbol{\sigma}') = \nabla \cdot (\mathbf{u}' \cdot \boldsymbol{\sigma}). \tag{31}$$

Integrating both sides over the volume $\mathcal{V}$ and applying the divergence theorem, we obtain Eq. 15.

2.4 Boundary integral formulation of the Laplace equation

Even with the aid of the Lorentz reciprocal theorem, it is necessary to solve the Stokes and (for self-phoretic particles) Laplace equations in a fluid domain containing the active particles as interior boundaries. For most configurations of the suspension, an analytical solution is intractable, and a numerical approach is required. Many numerical methods (e.g., the Finite Element Method) discretize and solve the governing partial differential equations in the three-dimensional fluid domain. This can be computationally intensive. Moreover, if the domain is unbounded in one or more dimensions, the computational domain must be truncated. Typically, the computational domain must be large in order to accurately approximate an unbounded solution, and significant computational effort must be expended on calculating the flow, pressure, and concentration fields far away from the particles.

An alternative approach proceeds from the following insight: a linear boundary value problems can be reformulated as a boundary integral equation (BIE) on the domain boundaries [51, 60]. Furthermore, the boundary integral equation can be discretized for numerical solution, yielding a dense linear system of coupled *boundary element* equations in the form of $\mathbf{A} \cdot \mathbf{q} = \mathbf{b}$. One significant advantage of the *boundary element method* is that it requires discretization of only bounding surfaces; for instance, to represent an unbounded system of N particles, one need only mesh the surfaces of the N spheres. As a disadvantage, the coefficient matrix $\mathbf{A}$ is typically fully populated and non-symmetric; therefore, for a system of N_{elm} elements, the required computer memory scales as $\mathcal{O}(N_{\mathrm{elm}}^2)$, and the required computation time scales as $\mathcal{O}(N_{\mathrm{elm}}^3)$.

In order to obtain the BIE for the Laplace equation, we first consider the divergence theorem:

$$\int_{\mathcal{V}} \nabla \cdot \mathbf{A} dV = -\int_{\mathcal{S}} \mathbf{A} \cdot \hat{\mathbf{n}} \; dS, \tag{32}$$

where the volume integral on the left hand side is carried out over the entire solution domain $\mathcal{V}$, and the surface integral on the right hand side is carried out over all boundaries $\mathcal{S}$. We include a negative sign on the right hand side of the equation because we define $\hat{\mathbf{n}}$ to point *into* the solution domain (see **Figure** 3). If $\mathbf{A} = \phi\nabla\psi$, where $\phi(\mathbf{x})$ and $\psi(\mathbf{x})$ are scalar fields, then the divergence term $\nabla \cdot \mathbf{A} = \phi\nabla^2\psi + \nabla\phi \cdot \nabla\psi$, and we obtain Green's first identity:

$$\int_{\mathcal{V}} \left(\phi\nabla^2\psi + \nabla\phi \cdot \nabla\psi\right) dV = -\int_{\mathcal{S}} \phi\nabla\psi \cdot \hat{\mathbf{n}} \; dS. \tag{33}$$

We can also write Green's first identity for $\mathbf{A} = \psi\nabla\phi$:

$$\int_{\mathcal{V}} \left(\psi\nabla^2\phi + \nabla\psi \cdot \nabla\phi\right) dV = -\int_{\mathcal{S}} \psi\nabla\phi \cdot \hat{\mathbf{n}} \; dS. \tag{34}$$

Subtracting Eq. 34 from Eq. 33, we obtain Green's second identity:

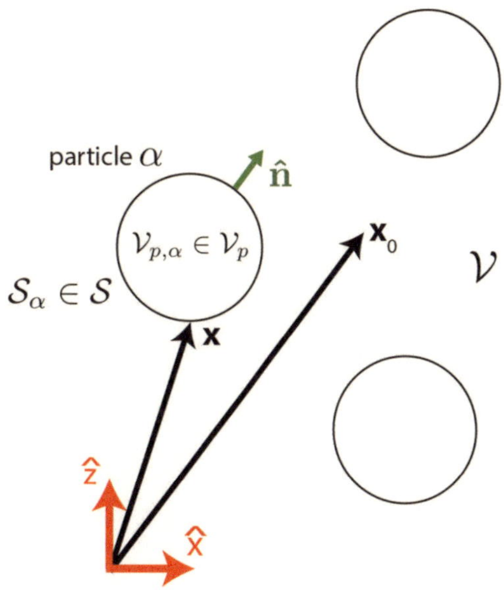

Figure 3.
Schematic illustration of the geometry for development of the boundary integral equations for the Laplace and stokes equations. The fluid domain is denoted by $\mathcal{V}$, the solid domain by $\mathcal{V}_p$, and the interior of particle α by $\mathcal{V}_{p,\alpha}$, where $\mathcal{V}_p = \cup_{\alpha=1}^N \mathcal{V}_{p,\alpha}$. The solid and fluid domains are separated by the particle surfaces $\mathcal{S}_\alpha$, with $\mathcal{S} = \cup_{\alpha=1}^N \mathcal{S}_\alpha$. The observation point $\boldsymbol{x}_0$ can occur in $\mathcal{V}$, in $\mathcal{V}$, or on $\mathcal{S}$; we show $\boldsymbol{x}_0$ in $\mathcal{V}$.

$$\int_{\mathcal{V}} (\phi \nabla^2 \psi - \psi \nabla^2 \phi) \, dV = -\int_{S} (\phi \nabla \psi - \psi \nabla \phi) \cdot \hat{\mathbf{n}} \ dS. \tag{35}$$

Now, we let $\phi = c(\mathbf{x})$, with $\nabla^2 c = 0$. Furthermore, we let $\psi = G(\mathbf{x}, \mathbf{x}_0)$, where the Green's function $G(\mathbf{x}, \mathbf{x}_0)$ satisfies Poisson's equation:

$$\nabla^2 G(\mathbf{x}, \mathbf{x}_0) + \delta(\mathbf{x} - \mathbf{x}_0) = 0. \tag{36}$$

We obtain:

$$\int_{\mathcal{V}} c(\mathbf{x}) \, \nabla^2 G(\mathbf{x}, \mathbf{x}_0) dV = -\int_{S} [c(\mathbf{x}) \nabla G(\mathbf{x}, \mathbf{x}_0) - G(\mathbf{x}, \mathbf{x}_0) \nabla c(\mathbf{x})] \cdot \hat{\mathbf{n}} \ dS. \tag{37}$$

We have not yet specified the location of the pole $\mathbf{x}_0$. If $\mathbf{x}_0$ is located in the domain $\mathcal{V}$, then, using the properties of the Dirac delta function, we obtain an integral representation of the concentration field $c(\mathbf{x}_0)$ at a point $\mathbf{x}_0 \in \mathcal{V}$:

$$c(\mathbf{x}_0) = \int_{S} [c(\mathbf{x}) \nabla G(\mathbf{x}, \mathbf{x}_0) - G(\mathbf{x}, \mathbf{x}_0) \nabla c(\mathbf{x})] \cdot \hat{\mathbf{n}} \ dS. \tag{38}$$

Using the divergence theorem, can show that $\nabla^2 \left(\frac{1}{|\mathbf{x} - \mathbf{x}_0|} \right) = -4\pi \delta(\mathbf{x} - \mathbf{x}_0)$, so that

$$G(\mathbf{x}, \mathbf{x}_0) = \frac{1}{4\pi |\mathbf{x} - \mathbf{x}_0|}. \tag{39}$$

We recall from electrostatics that $G(\mathbf{x}, \mathbf{x}_0)$ represents the electrostatic potential at $\mathbf{x}$ from a point charge of unit strength located at $\mathbf{x}_0$. Within the context of the steady diffusion equation $\nabla^2 c = 0$, it has a different physical interpretation: it can be regarded as the steady concentration $c(\mathbf{x})$ at a point $\mathbf{x}$ due to a point-like, steady source of concentration, continuously injected into the system, located at $\mathbf{x}_0$ and with unit strength (i.e., unit number density flux per unit time). One can take derivatives of $G(\mathbf{x}, \mathbf{x}_0)$ with respect to $\mathbf{x}_0$ to obtain higher order multipole singularities. For instance, we can obtain the Green's function $\mathbf{G}_{\mathrm{dp}}(\mathbf{x}, \mathbf{x}_0)$ for a source/sink dipole located at $\mathbf{x}_0$ as

$$\mathbf{G}_{\mathrm{dp}}(\mathbf{x}, \mathbf{x}_0) \equiv \nabla_{\mathbf{x}_0} G(\mathbf{x}, \mathbf{x}_0) = \frac{(\mathbf{x} - \mathbf{x}_0)}{4\pi |\mathbf{x} - \mathbf{x}_0|^3}. \tag{40}$$

As $\mathbf{G}_{\mathrm{dp}}(\mathbf{x}, \mathbf{x}_0)$ is a vector quantity, we obtain a scalar contribution to $c(\mathbf{x})$ by multiplying with a dipole vector $\mathbf{d}$; the magnitude and direction of $\mathbf{d}$ specify the strength and orientation of the dipole.

By inspection, the Green's function obeys the symmetry property $G(\mathbf{x}, \mathbf{x}_0) = G(\mathbf{x}_0, \mathbf{x})$, so we can rewrite Eq. 38 as

$$c(\mathbf{x}_0) = \int_{S} [c(\mathbf{x}) \nabla G(\mathbf{x}_0, \mathbf{x}) - G(\mathbf{x}_0, \mathbf{x}) \nabla c(\mathbf{x})] \cdot \hat{\mathbf{n}} \ dS. \tag{41}$$

Interestingly, we have obtained an expression for $c(\mathbf{x}_0)$ in the solution domain in terms of the values of $c(\mathbf{x})$ and $\nabla c(\mathbf{x}) \cdot \hat{\mathbf{n}}$ on the domain boundaries. Note that this is not a solution to a boundary value problem for $c(\mathbf{x}_0)$, since a BVP specifies only one of the quantities $c(\mathbf{x})$ and $\nabla c(\mathbf{x}) \cdot \hat{\mathbf{n}}$ on the domain boundaries. Specifically, for the problem of a system of catalytic particles outlined above, we only know $\nabla c(\mathbf{x}) \cdot \hat{\mathbf{n}}$

a priori. Eq. 41 has an interesting physical interpretation: $c(\mathbf{x}_0)$ can be regarded as the concentration due to a distribution of monopoles (i.e., point sources of mass flux) with strength $-\nabla c(\mathbf{x}) \cdot \hat{\mathbf{n}}$ on the boundaries, *plus* a distribution of point dipoles (i.e., infinitesimally separated pairs of mass sources and sinks) with strength $c(\mathbf{x})$ and orientation $\hat{\mathbf{n}}$ on the boundaries.

We still have two other options for where to place $\mathbf{x}_0$: inside the boundary $\mathcal{S}$ (i.e., outside the solution domain $\mathcal{V}$) or somewhere on the boundary $\mathcal{S}$. If we place $\mathbf{x}_0$ inside $\mathcal{S}$, then the integral on the left hand side of Eq. 37 vanishes:

$$\int_{\mathcal{S}} [c(\mathbf{x})\nabla G(\mathbf{x}, \mathbf{x}_0) - G(\mathbf{x}, \mathbf{x}_0)\nabla c(\mathbf{x})] \cdot \hat{\mathbf{n}} \; dS = 0. \tag{42}$$

Placing $\mathbf{x}_0$ on the boundary requires some care in how to handle the Dirac delta function on the left hand side of Eq. 37. If we regard the Dirac delta as the limit of a sequence of distributions, then it is clear that a factor of one half should arise when we integrate over $\mathcal{V}$:

$$\frac{1}{2}c(\mathbf{x}_0) = \int_{\mathcal{S}} [c(\mathbf{x})\nabla G(\mathbf{x}, \mathbf{x}_0) - G(\mathbf{x}, \mathbf{x}_0)\nabla c(\mathbf{x})] \cdot \hat{\mathbf{n}} \; dS. \tag{43}$$

This is a *boundary integral equation* (BIE) because the left hand side is the concentration $c(\mathbf{x}_0)$ at a point on the boundary, while the right hand side is an integral of $c(\mathbf{x})$ and $\nabla c(\mathbf{x}) \cdot \hat{\mathbf{n}}$ over the same boundary. A boundary value problem typically specifies one of these two quantities on the boundary; the other can be obtained with Eq. 43.

In the *boundary element method*, the boundary integral equation is discretized for numerical solution. Here, we briefly summarize the method, and direct the reader to consult the useful and comprehensive book of Pozrikidis for further information [51]. Each particle is represented as a meshed, closed surface. The meshing only needs to be done once; for a dynamical simulation, no remeshing during the simulation is required, even if the particles move relative to each other. The concentration c and its normal derivative $\nabla c \cdot \hat{\mathbf{n}}$ are assumed to be uniform over element i. For a point $\mathbf{x}_0$ on the surface, we obtain the boundary integral equation:

$$\frac{1}{2}c(\mathbf{x}_0) = \sum_{i=1}^{N_{\text{elm}}} \left[c_i \left(\int_{\mathcal{S}_i} \nabla G(\mathbf{x}, \mathbf{x}_0) \cdot \hat{\mathbf{n}} \; dS \right) - (\nabla c \cdot \hat{\mathbf{n}})_i \left(\int_{\mathcal{S}_i} G(\mathbf{x}, \mathbf{x}_0) \; dS \right) \right]. \tag{44}$$

Choosing $\mathbf{x}_0$ as the midpoint $\mathbf{x}_j$ of element j, we can write N_{elm} equations:

$$\frac{1}{2}c_j = \sum_{i=1}^{N_{\text{elm}}} \left[c_i \left(\int_{\mathcal{S}_i} \nabla G(\mathbf{x}, \mathbf{x}_j) \cdot \hat{\mathbf{n}} \; dS \right) - (\nabla c \cdot \hat{\mathbf{n}})_i \left(\int_{\mathcal{S}_i} G(\mathbf{x}, \mathbf{x}_j) \; dS \right) \right]. \tag{45}$$

The N_{elm} equations can be written in matrix form:

$$\left(A_{ij} - \frac{1}{2}\delta_{ij} \right) c_j = B_{ij} (\nabla c \cdot \hat{\mathbf{n}})_j, \tag{46}$$

where

$$A_{ij} \equiv \int_{\mathcal{S}_i} \nabla G(\mathbf{x}, \mathbf{x}_j) \cdot \hat{\mathbf{n}} \; dS \tag{47}$$

and

$$B_{ij} \equiv \int_{S_i} G(\mathbf{x}, \mathbf{x}_j) \; dS. \tag{48}$$

Given either a specification of either c_j (Dirichlet boundary conditions) or $(\nabla c \cdot \hat{\mathbf{n}})_j$ (Neumann boundary conditions), the algebraic system in Eq. 46 can be solved numerically with standard methods.

A certain difficulty becomes apparent when we consider the element B_{ii}: the evaluation point $\mathbf{x}_i$ lies within the element of integration, and therefore the integral contains the singularity of Eq. 39. We are saved from a potentially disastrous situation by the fact that the integral is carried out over an area. Nevertheless, this singular integral has to be handled with care. Further technical information, as well as a wealth of practical details concerning implementation of the BEM, is available in Ref. [51].

As a further note, issues with singular integrals have motivated development of *regularized boundary element methods*, which use a regularized Green's function, i.e., a Green's function with the singularity "smeared out" over a finite size ε [52, 54, 61].

2.5 Boundary integral formulation of the Stokes equation

A similar approach can be taken for the Stokes equation [51, 59]. Recall that the Stokes equation is:

$$\nabla \cdot \boldsymbol{\sigma} = -\nabla P + \mu \nabla^2 \mathbf{u} = 0. \tag{49}$$

We can define a Green's function $\mathcal{G}(\mathbf{x}, \mathbf{x}_0)$ as the solution $\mathbf{u}(\mathbf{x}) \equiv \mathcal{G}(\mathbf{x}, \mathbf{x_0}) \cdot \mathbf{F}$ to the Stokes equation with a body point force $\mathbf{F}$ located at $\mathbf{x}_0$:

$$-\nabla P + \mu \nabla^2 \mathbf{u} + \mathbf{F}\delta(\mathbf{x} - \mathbf{x}_0) = 0, \tag{50}$$

or

$$\nabla \cdot \boldsymbol{\sigma} = -\mathbf{F}\delta(\mathbf{x} - \mathbf{x}_0). \tag{51}$$

It can be shown that the Green's function is

$$\mathcal{G}_{ij}(\mathbf{x}, \mathbf{x}_0) = \frac{1}{8\pi\mu r}\left(\delta_{ij} + \frac{\tilde{x}_i \tilde{x}_j}{r^2}\right), \tag{52}$$

where $r \equiv |\mathbf{x} - \mathbf{x}_0|$ and where $\tilde{x}_j \equiv x_j - x_{0,j}$. Eq. 52 is commonly called the Oseen tensor or "Stokeslet". The fluid pressure in response to the point force is given by $P(\mathbf{x}) \equiv \mathcal{P}(\mathbf{x}, \mathbf{x}_0) \cdot \mathbf{F}$, where

$$\mathcal{P}_j = \frac{\tilde{x}_j}{4\pi r^3}. \tag{53}$$

The stress in the fluid is given by $\boldsymbol{\sigma} \equiv \Sigma(\mathbf{x}, \mathbf{x}_0) \cdot \mathbf{F}$, where

$$\Sigma_{ijk} = -3\frac{\tilde{x}_i \tilde{x}_j \tilde{x}_k}{4\pi r^5}. \tag{54}$$

Now we wish to apply Eq. 30. We specify the "primed" fields $\mathbf{u}'$ and $\boldsymbol{\sigma}'$ as the fields due to a point force $\mathbf{F}$ at $\mathbf{x}_0$ in unbounded fluid. For the "unprimed" fields $\mathbf{u}$

and $\boldsymbol{\sigma}$, we specify that they are fields of interest in the domain $\mathcal{V}$ bounded by $\mathcal{S}$ (see **Figure** 3). Furthermore, $\mathcal{V}$ is free of body forces, so that $\nabla \cdot \boldsymbol{\sigma} = 0$. We obtain:

$$\frac{\partial}{\partial x_j}\left(u_i \Sigma_{ijk} F_k\right) - u_i \frac{\partial}{\partial x_j}\left(\Sigma_{ijk} F_k\right) = \frac{\partial}{\partial x_j}\left(\sigma_{ij} \mathcal{G}_{ik} F_k\right) \tag{55}$$

$$F_k \frac{\partial}{\partial x_j}\left(u_i \Sigma_{ijk}\right) + u_i F_i \delta(\mathbf{x} - \mathbf{x}_0) = F_k \frac{\partial}{\partial x_j}\left(\mathcal{G}_{ik} \sigma_{ij}\right) \tag{56}$$

We integrate both sides over the domain $\mathcal{V}$:

$$\int_{\mathcal{V}} u_k F_k \delta(\mathbf{x} - \mathbf{x}_0) dV = F_k \int \left[\frac{\partial}{\partial x_j}\left(\mathcal{G}_{ik}\sigma_{ij}\right) - \frac{\partial}{\partial x_j}\left(u_i \Sigma_{ijk}\right)\right] dV \tag{57}$$

Now we apply the divergence theorem:

$$F_k \int_{\mathcal{V}} u_k \delta(\mathbf{x} - \mathbf{x}_0) dV = -F_k \int_{\mathcal{S}} \left[\mathcal{G}_{ik}(\mathbf{x}, \mathbf{x}_0)\sigma_{ij} - u_i \Sigma_{ijk}(\mathbf{x}, \mathbf{x}_0)\right] n_j dS, \tag{58}$$

where the negative sign appears because of our convention that $\hat{\mathbf{n}}$ points into $\mathcal{V}$. We note that $\mathcal{G}_{ik}(\mathbf{x}, \mathbf{x}_0) = \mathcal{G}_{ik}(\mathbf{x}_0, \mathbf{x})$ and $\Sigma_{ijk}(\mathbf{x}, \mathbf{x}_0) = -\Sigma_{ijk}(\mathbf{x}_0, \mathbf{x})$. Additionally, the choice of F_k was arbitrary. We can therefore write:

$$\int_{\mathcal{V}} u_k \delta(\mathbf{x} - \mathbf{x}_0) dV = -\int_{\mathcal{S}} \left[\mathcal{G}_{ik}(\mathbf{x}_0, \mathbf{x})\sigma_{ij} + u_i \Sigma_{ijk}(\mathbf{x}_0, \mathbf{x})\right] n_j dS. \tag{59}$$

If we choose to place $\mathbf{x}_0$ in $\mathcal{V}$, we obtain a boundary integral representation for $u_k(\mathbf{x}_0)$:

$$u_k(\mathbf{x}_0) = -\int_{\mathcal{S}} \left[\mathcal{G}_{ik}(\mathbf{x}_0, \mathbf{x})\sigma_{ij}(\mathbf{x}) + \Sigma_{ijk}(\mathbf{x}_0, \mathbf{x}) u_i(\mathbf{x})\right] n_j dS. \tag{60}$$

As with Eq. 41, the boundary integral representation for the flow field has an interesting physical interpretation. The first term on the right hand side of Eq. 60 can be regarded as a "single layer potential" due to a distribution of point forces with strength $\boldsymbol{\sigma} \cdot \hat{\mathbf{n}}$ over the surface of the particle. The second term on the right hand side is the "double layer potential." Detailed examination of this term reveals that it can be decomposed into the superposition of the flow due to a distribution $\mathbf{u} \cdot \hat{\mathbf{n}}$ of point sources and sinks of fluid mass, plus the flow to a distribution of point force dipoles [59].

If we place $\mathbf{x}_0$ outside $\mathcal{V}$, i.e., inside the space $\mathcal{V}_p$ enclosed by the particles, we obtain

$$-\int_{\mathcal{S}} \left[\mathcal{G}_{ik}(\mathbf{x}_0, \mathbf{x})\sigma_{ij}(\mathbf{x}) + \Sigma_{ijk}(\mathbf{x}_0, \mathbf{x}) u_i(\mathbf{x})\right] n_j dS = 0. \tag{61}$$

Finally, if we place $\mathbf{x}_0$ on the boundary $\mathcal{S}$, we obtain a boundary integral equation:

$$\frac{1}{2} u_k(\mathbf{x}_0) = -\int_{\mathcal{S}} \left[\mathcal{G}_{ik}(\mathbf{x}_0, \mathbf{x})\sigma_{ij}(\mathbf{x}) + \Sigma_{ijk}(\mathbf{x}_0, \mathbf{x}) u_i(\mathbf{x})\right] n_j dS. \tag{62}$$

For rigid body motion, *including* the 6N "primed" problems for the Lorentz reciprocal theorem, the double layer can be eliminated from the boundary integral equation as follows [59]. Consider extending the volume filled by "fluid" to $\mathcal{V}_p$. Within $\mathcal{V}_p$, the flow field $\mathbf{u}$ is simply the flow field $\mathbf{u}^{\mathrm{RBM}}$ for rigid body motion, since it must obey no-slip on the surface $\mathcal{S}$. Now we apply Eq. 59 for the field $\mathbf{u}^{\mathrm{RBM}}$ inside $\mathcal{V}_p$ and $\mathbf{x}_0 \in \mathcal{V}$, noting that we must use a normal $\hat{\mathbf{n}}' = -\hat{\mathbf{n}}$ pointing *into* $\mathcal{V}_p$:

$$-\int_{\mathcal{S}} \left[\mathcal{G}_{ik}(\mathbf{x}_0, \mathbf{x}) \sigma_{ij}^{\mathrm{RBM}}(\mathbf{x}) + \Sigma_{ijk}(\mathbf{x}_0, \mathbf{x}) u_i^{\mathrm{RBM}}(\mathbf{x}) \right] n_j' dS = 0. \tag{63}$$

For rigid body motion, there is no shear stress and the pressure is uniform, i.e., $\sigma_{ij}^{\mathrm{RBM}}(\mathbf{x})\hat{\mathbf{n}}' = -p_0 \hat{\mathbf{n}}'$. The first term is simply the integral of $\mathcal{G}_{ik}(\mathbf{x}_0, \mathbf{x}) \cdot \hat{\mathbf{n}}'$ over the surface $\mathcal{S}$ for $\mathbf{x}_0 \in \mathcal{V}$, which vanishes identically by incompressibility. This leaves:

$$-\int_{\mathcal{S}} \left[\Sigma_{ijk}(\mathbf{x}_0, \mathbf{x}) u_i^{\mathrm{RBM}}(\mathbf{x}) \right] n_j' dS = 0. \tag{64}$$

Examining Eq. 62, we note that $\mathbf{u}(\mathbf{x})$, i.e., the flow velocity in $\mathcal{V}$, is equal to $\mathbf{u}^{\mathrm{RBM}}$ on $\mathcal{S}$. Therefore, we conclude:

$$u_k(\mathbf{x}_0) = -\int_{\mathcal{S}} \left[\mathcal{G}_{ik}(\mathbf{x}_0, \mathbf{x}) \sigma_{ij}(\mathbf{x}) \right] n_j dS, \quad \mathbf{x}_0 \in \mathcal{V}. \tag{65}$$

In order to obtain a single-layer boundary integral equation for $\mathbf{x}_0 \in \mathcal{S}$, note that the jump discontinuity responsible for the factor of 1/2 in Eq. 62 is strictly from the double-layer potential [59]. The contribution of the single layer potential to the velocity field is continuous across $\mathcal{S}$. We obtain:

$$u_k(\mathbf{x}_0) = -\int_{\mathcal{S}} \left[\mathcal{G}_{ik}(\mathbf{x}_0, \mathbf{x}) \sigma_{ij}(\mathbf{x}) \right] n_j dS, \quad \mathbf{x}_0 \in \mathcal{S}. \tag{66}$$

This single layer boundary integral equation can be discretized and solved numerically in a similar manner as the Laplace equation; Ref. 51 provides a comprehensive account.

2.6 Active suspensions in confined geometries

In the preceding, we considered a suspension of N particles in an unbounded three-dimensional geometry. However, the presence of confining boundaries can have a significant effect on the dynamics of a suspension. It is possible to include a solid surface by explicitly meshing it and including it as a "fixed" or immobile particle in the calculations [53]. This approach is typically necessary for solid surfaces with corners or complex topography. One disadvantage of this approach is that an infinite surface (e.g., an infinite planar wall) must be truncated and included as a finite size object. Care must be taken that the mesh is sufficiently fine near the particles, so that, for instance, the concentration and flow fields do not "leak" through a solid wall, but also that the mesh is sufficiently coarse far away from the particles, so that computation time is tractable.

A second, "mesh-free" approach is suitable for confining geometries with high symmetry, such as an infinite planar wall [39], an interface between two fluids with different viscosities [62], a fluid domain bounded by a solid wall and a free interface [63], or even two infinite planar walls. Additionally, it can be suitable if the domain is periodic in two or three dimensions. In this approach, the Green's functions for the Laplace and Stokes equations are replaced with Green's functions that obey the desired boundary conditions on the bounding surfaces. The Green's function in the confined geometry can often be constructed by the method of images.

2.7 Thermal fluctuations

So far, we have considered the deterministic contributions to the $6N$ components of velocity for a suspension of N particles. However, as outlined in the Introduction, the interplay of these deterministic contributions and the stochastic Brownian forces on the particles is important—and in some problems, such as the long-time behavior of an active colloid, it is absolutely essential.

One approach to include Brownian forces on an active particle, the *hybrid boundary element/Brownian dynamics method*, simply calculates them separately and superposes them with the deterministic contributions. Using the Itô convention for stochastic differential equations, this superposition is expressed by the overdamped Langevin equation for the generalized, $6N$-component coordinate $\mathbf{q}$:

$$\frac{d\mathbf{q}}{dt} = \mathbf{V} + k_B T(\nabla \cdot \boldsymbol{\mathcal{M}}) + \sqrt{2k_B T}\mathbf{B} \cdot \mathbf{W}, \tag{67}$$

where $\mathbf{V}$ is the *deterministic* contribution of activity to the generalized velocity $(\mathbf{U}_\alpha, \boldsymbol{\Omega}_\alpha)^T$, i.e., the solution to the problem outlined above; $\boldsymbol{\mathcal{M}}$ is the grand mobility matrix $\boldsymbol{\mathcal{M}} = \boldsymbol{\mathcal{R}}^{-1}$; $\mathbf{B}$ satisfies $\mathbf{B} \cdot \mathbf{B}^T = \boldsymbol{\mathcal{M}}$; and $\mathbf{W}$ is a collection of independent Wiener processes. Discretizing time in steps of Δt, one can write a generalized displacement $\Delta\mathbf{q}$ as the following Euler-Maruyama equation [34, 64]:

$$\Delta\mathbf{q} = \mathbf{V}\Delta t + k_B T(\nabla \cdot \boldsymbol{\mathcal{M}})\Delta t + \sqrt{2k_B T}\mathbf{B} \cdot \Delta\mathbf{w}, \tag{68}$$

where $\Delta\mathbf{w}$ is a stochastic variable with $\langle \Delta\mathbf{w} \rangle = 0$ and $\langle \Delta w_i \Delta w_j \rangle = \delta_{ij}\Delta t$. The stochastic drift term $(\nabla \cdot \boldsymbol{\mathcal{M}})$ is a consequence of having a configuration-dependent mobility tensor in the framework of the Itô interpretation.

The update of the orientation of each particle α should respect the constraint that $|\mathbf{d}_\alpha| = 1$ and avoid any errors arising from application of (non-commuting) rotation matrices in arbitrary order to $\mathbf{d}_\alpha$. There are robust algorithms for rigid body motion that represent the orientations of the particles with quaternions [65], Euler angles [66], or rotation matrices that transform between body-fixed and global reference frames [67].

The stochastic drift term in Eq. 68 can present some difficulty for numerical calculations [66]. For some simple situations, such as a single spherical colloid near a planar wall [34, 42], solutions for the configuration dependence of the mobility tensor are available in the literature [68, 69]. Alternatively, Eq. 67 can be discretized and solved via Fixman's midpoint method to avoid calculation of the drift term [70].

This approach assumes that that the colloid and the fluid are not fluctuating on the same timescale, i.e., the fluid velocity is integrated out as a fast variable. Additionally, for self-phoretic particles, this approach necessarily neglects fluctuations of the chemical field $c(\mathbf{x})$ in the fluid domain $\mathcal{V}$.

A recently developed variation of the boundary element method for Stokes flow, the *fluctuating boundary element method*, does not make this *post hoc* superposition of deterministic and Brownian contributions to particle motion. Rather, fluctuations are directly incorporated into boundary integral equation via a random velocity field on the boundary $\mathcal{S}$ [71].

3. Discussion and conclusions

The boundary element method is emerging as a powerful and important method for numerical simulation in the field of synthetic active colloids [30, 52, 54–57]. This new area of application follows many years of fruitful application to modeling biological microswimmers, including with the squirmer model [40, 53]. For active colloids, a major advantage of the boundary element approach is that it can resolve the microscopic details of phoretic self-propulsion, including the chemical and flow fields generated by an active colloid, the surface chemistry and shape of the colloid, and the microscopic physics of how the colloid can couple to ambient fields and confining surfaces.

A few examples serve to illustrate the utility of the approach. Ref. [30] considers the dynamics of a spherical active Janus colloid near a planar wall. The colloid can "sense" and respond to the wall through self-generated chemical and hydrodynamic fields. Specifically, the wall provides a no-flux boundary condition for the solute concentration, and a no-slip boundary condition for the flow field. By confining the solute, the wall enriches the concentration of solute in the space between the particle and the wall, breaking the axial symmetry of the concentration field. Concerning the flow, the flow created by the particle scatters off the wall and back to the particle. These effects are captured by the boundary element method, including their dependence on the size of the catalytic cap and the spatial variation in the surface mobility b over the surface of the particle. As another example, Ref. [43] considers the dynamics of a *photo-active* spherical Janus colloid. The catalytic cap of the colloid is only active when exposed to incident light. This self-shadowing effect, in conjunction with the spatial variation of b on the surface of the colloid, leads to phototaxis (rotation of the cap towards the light) or anti-phototaxis (rotation of the cap away from the light.) Notably, this work uses the hybrid BEM/BD method to calculate the distribution of particle orientations as a function of illumination intensity and particle surface chemistry. Concerning the interaction of multiple particles, Ref. 57 uses the regularized BEM to calculate the dynamics of multiple isotropic spherical colloids. Interestingly, a group of $N = 5$ particles can form a stable cluster with broken rotational symmetry. This broken symmetry allows propulsion of the whole cluster. Finally, concerning shape, the BEM has been used to model toroidal [54] and spherocylindrical [72] self-phoretic particles.

However, some caveats are in order. For the hybrid boundary element/Brownian dynamics method discussed in this work, neither the fluctuations of the suspending fluid nor of the chemical field(s) are explicitly resolved. For self-phoretic particles in the ångstrom to nanometer size range, the particles, the solute, and the solvent fluctuate on similar timescales. Additionally, the validity of the continuum description of the surrounding solution is questionable. Molecular and mesoscopic simulation methods that resolve discrete solute and solvent particles may be more appropriate in this size range [48]. As a second caveat, boundary element methods are most suited to solution of linear governing PDEs, such as the Laplace and Stokes equations. Introducing nonlinearity in the governing equations (e.g., for a solution with nonlinear rheology or nonlinear bulk reaction kinetics) leads to the appearance of volume integrals in the boundary integral formulation. Thirdly and relatedly, the boundary element method is not as easily extensible as other methods (e.g., the finite element method) for inclusion of more complicated multiphysics. Finally, there is a caveat specific to active colloids. Much remains unknown about the reaction kinetics for self-phoretic particles. The boundary element method can have many free microscopic parameters (e.g., the values of the surface mobility b on different surfaces); this raises the danger of overfitting to experimental results.

As a potential direction of research, we suggest developing a hybrid computational method combining the advantages of BEM and Stokesian Dynamics (SD). Stokesian dynamics is a method for simulating the dynamics of colloidal suspensions [73–76]. Far-field hydrodynamic interactions are included in SD, truncated at the level of the stresslet (i.e., the first moment of the stress on the surface of a particle, which produces a hydrodynamic disturbance decaying as $\sim 1/r^2$.) Near-field hydrodynamic interactions are typically included via lubrication forces acting between particle pairs. Due to these approximations, Stokesian dynamics is computationally much cheaper than BEM, allowing access to collective dynamics, the rheology of dense suspensions, etc. On the other hand, SD does not typically resolve the microscopic details of individual particles, such as shape or heterogeneous surface chemistry. A hybrid BEM-SD method could combine the detailed microscopic resolution of BEM for near-field interactions with the ability of SD to capture many-body phenomena driven by far-field interactions. (This hybrid approach would bear some similarity to the fast multipole method.) In the Appendix at the end of this chapter, we develop a starting point for including interfacial flows $\mathbf{v}_s(\mathbf{x})$ within the standard SD formalism for spherical particles.

As a second potential research direction, one could consider *deformable* active particles using the BEM. The boundary element method for Stokes flow has been coupled to methods to model particle elasticity, including the finite element method, in order to study the deformation of fluid-filled capsules [77] and elastic particles in shear flow [78], as well as the deformation of blood cells squeezing through constrictions [79].

The boundary element method could also be used to investigate questions touching upon fundamental nonequilibrium statistical mechanics. For instance, do nonequilibrium steady states of squirmers or self-phoretic particles (e.g., stable clusters of catalytic particles [57]) minimize the rate of entropy production [80]? When do hydrodynamic interactions suppress or enhance motility-induced phase separation and other nonequilibrium phase transitions? Does the pressure of an active suspension on a boundary obey an equation of state when hydrodynamic and phoretic interactions with the boundary are considered [76, 81]?

In any case, we anticipate that the boundary element method will continue to find successful application in the microswimmers field. A few potential problems include: modeling the collision dynamics and scattering of two or more non-spherical active colloids [72, 82]; the interaction of an active colloid and a passive colloid, possibly including the formation of dimeric bound states for cargo transport; and further exploration of motion near bounding surfaces and interfaces, especially fluid/fluid interfaces.

A. Faxén laws and connection to Stokesian dynamics

Consider an inert (non-active) sphere of radius R in an ambient flow field $\mathbf{u}^\infty(\mathbf{x})$. The sphere has translational velocity $\mathbf{U}$ and rotational velocity $\mathbf{\Omega}$. The flow field $\mathbf{u}$ can be written as $\mathbf{u}(\mathbf{x}) = \mathbf{u}^\infty(\mathbf{x}) + \mathbf{u}^D(\mathbf{x})$, where $\mathbf{u}^D(\mathbf{x})$ is the velocity disturbance created by the presence of the sphere. The boundary condition for $\mathbf{u}^D(\mathbf{x})$ on the sphere surface $\mathcal{S}$ is

$$\mathbf{u}^D(\mathbf{x_s}) = \mathbf{U} + \mathbf{\Omega} \times (\mathbf{x_s} - \mathbf{x}_0) - \mathbf{u}^\infty(\mathbf{x}_s), \quad \mathbf{x}_s \in \mathcal{S}. \tag{69}$$

Additionally, $\mathbf{u}^D(|\mathbf{x}| \to \infty) \to 0$. Taking the sphere position to be $\mathbf{x}_0 = 0$, we can expand the ambient flow field around the sphere center as:

$$u_i^\infty(\mathbf{x}) = u_i^\infty(0) + \left.\frac{\partial u_i^\infty}{\partial x_j}\right|_{\mathbf{x}=0} x_j + \frac{1}{2}\left.\frac{\partial^2 u_i^\infty}{\partial x_j \partial x_k}\right|_{\mathbf{x}=0} x_j x_k + \ldots \tag{70}$$

Now we recall the definitions of the (symmetric) rate of strain tensor e_{ij},

$$e_{ij} = \frac{1}{2}\left(\frac{\partial u_i}{\partial x_j} + \frac{\partial u_j}{\partial x_i}\right), \tag{71}$$

and the (anti-symmetric) vorticity tensor

$$W_{ij} = \frac{1}{2}\left(\frac{\partial u_i}{\partial x_j} - \frac{\partial u_j}{\partial x_i}\right), \tag{72}$$

The vorticity tensor can be related to the vorticity *vector* $\mathbf{w} = \nabla \times \mathbf{u}$ by

$$\mathbf{W} = \frac{1}{2}\boldsymbol{\varepsilon} \cdot \mathbf{w}. \tag{73}$$

Here, $\boldsymbol{\varepsilon}$ is the Levi-Civita tensor. The first derivative in Eq. 70 can be decomposed into symmetric and anti-symmetric contributions:

$$\frac{\partial u_i^\infty}{\partial x_j} = e_{ij}^\infty + W_{ij}^\infty. \tag{74}$$

Using the Lorentz reciprocal theorem, one can obtain Faxén's law for the drag force on the sphere (see Ref. [59] for details):

$$\mathbf{F}^{\text{drag}} = 6\pi\mu R\left[\left(1 + \frac{R^2}{6}\nabla^2\right)\mathbf{u}^\infty(\mathbf{x}_0) - \mathbf{U}\right]. \tag{75}$$

(In our shorthand notation, the Laplacian is first applied to $\mathbf{u}^\infty(\mathbf{x})$ and then evaluated at $\mathbf{x}_0$.) One can also obtain Faxén law for the drag torque:

$$\tau^{\text{drag}} = 8\pi\mu R^3(\boldsymbol{\omega}^\infty(\mathbf{x}_0) - \boldsymbol{\Omega}), \tag{76}$$

where the angular velocity of the fluid $\boldsymbol{\omega} \equiv \frac{1}{2}\mathbf{w}$. Finally, there is a Faxén law for the stresslet [59, 83, 84]

$$\mathbf{S} = \frac{20}{3}\pi\mu R^3\left(1 + \frac{R^2}{10}\nabla^2\right)\mathbf{e}^\infty(\mathbf{x}_0), \tag{77}$$

where $\mathbf{S}$ is defined as an integral over the particle surface

$$S_{ij} = \int\left[\frac{1}{2}\left(x_j\sigma_{ik}n_k + x_i\sigma_{jk}n_k\right) - \frac{1}{3}(x_k\sigma_{kl}n_l)\delta_{ij} - \mu\left(u_i n_j + u_j n_i\right)\right]dS. \tag{78}$$

So far we have only presented standard results, but now we raise the following question. Consider an *active* sphere with a slip velocity $\mathbf{v}_s(\mathbf{x})$. Comparing the boundary conditions in Eq. 3 and Eq. 69, can we construct an ambient *linear* flow field

$$\mathbf{u}^\infty(\mathbf{x}) = \mathbf{u}^\infty(0) + \mathbf{e}^\infty \cdot \mathbf{x} + \boldsymbol{\omega}^\infty \times \mathbf{x} \tag{79}$$

with $\mathbf{u}^\infty(\mathbf{x}_s) = -\mathbf{v}_s(\mathbf{x}_s)$ on $\mathcal{S}$? Constructing an effective flow field would allow us to obtain $\mathbf{F}^{\text{drag}}$, τ^{drag}, and $\mathbf{S}$ by Faxén laws, without having to solve the complete hydrodynamic problem posed in Section II. Moreover, an understanding of how $\mathbf{v}_s(\mathbf{x})$ determines $\mathbf{F}^{\text{drag}}$, τ^{drag}, and $\mathbf{S}$ would pave the way towards development of a hybrid BEM-Stokesian Dynamics scheme, since these quantities are central to SD.

As our starting point, we write the Taylor expansion of $\mathbf{u}^\infty(\mathbf{x})$:

$$u_i^\infty(\mathbf{x}) = u_i^\infty(0) + \left.\frac{\partial u_i^\infty}{\partial x_j}\right|_{\mathbf{x}=0} x_j. \tag{80}$$

To obtain $u_i^\infty(0)$, we integrate both sides of Eq. 80 over the surface of the sphere:

$$\int u_i^\infty(\mathbf{x})dS = \int u_i^\infty(0)dS + \int \left.\frac{\partial u_i^\infty}{\partial x_j}\right|_{\mathbf{x}=0} x_j dS. \tag{81}$$

We identify $\mathbf{u}_i^\infty(\mathbf{x})$ on the surface of the sphere as $-\mathbf{v}_s(\mathbf{x}_s)$. The second integral on the right hand side of Eq. 81 vanishes, giving

$$\mathbf{u}^\infty(0) = -\frac{1}{4\pi R^2}\int \mathbf{v}^s(\mathbf{x}_s)dS. \tag{82}$$

Using Eq. 75, we obtain

$$\mathbf{F}^{\text{drag}} = 6\pi\mu R\left[-\frac{1}{4\pi R^2}\int \mathbf{v}^s(\mathbf{x}_s)dS - \mathbf{U}\right]. \tag{83}$$

If we consider a force-free swimmer, $F^{\text{drag}} = 0$, giving the result:

$$\mathbf{U} = -\frac{1}{4\pi R^2}\int \mathbf{v}^s(\mathbf{x}_s)dS. \tag{84}$$

This equation is one of the major results obtained in Ref. 37 by use of the Lorentz reciprocal theorem. However, our rederivation and interpretation in terms of an effective ambient flow field $\mathbf{u}^\infty$ is (to our knowledge) novel. To obtain the vorticity associated with $\mathbf{u}^\infty$, we multiply Eq. 80 by $\varepsilon_{lmi}x_m$ and integrate over the sphere surface:

$$\int \varepsilon_{lmi}x_m u_i^\infty(\mathbf{x})dS = \int u_i^\infty(0)\varepsilon_{lmi}x_m dS + \int \left.\frac{\partial u_i^\infty}{\partial x_j}\right|_{\mathbf{x}=0} \varepsilon_{lmi}x_m\ x_j dS. \tag{85}$$

The first integral on the right hand side of Eq. 85 vanishes. For the second integral on the right hand side, we use the identity

$$\int x_m x_j dS = \frac{4\pi R^4}{3}\delta_{jm}. \tag{86}$$

We obtain:

$$\int \varepsilon_{lmi}x_m u_i^\infty(\mathbf{x})dS = \frac{4\pi R^4}{3}\varepsilon_{lji}\left.\frac{\partial u_i^\infty}{\partial x_j}\right|_{\mathbf{x}=0} \tag{87}$$

$$-\int \mathbf{x}\times\mathbf{v}^s(\mathbf{x}_s)dS = \frac{4\pi R^4}{3}\nabla\times\mathbf{u}^\infty\Big|_{\mathbf{x}=0} = \frac{8\pi R^4}{3}\boldsymbol{\omega}^\infty(0), \tag{88}$$

so that

$$\omega^{\infty}(0) = -\frac{3}{8\pi R^4}\int \mathbf{x} \times \mathbf{v}^s(\mathbf{x}_s)\,dS. \tag{89}$$

Using Eq. 76, we obtain:

$$\boldsymbol{\tau}^{\mathrm{drag}} = 8\pi\mu R^3\left(-\frac{3}{8\pi R^4}\int \mathbf{x} \times \mathbf{v}^s(\mathbf{x}_s)\,dS - \boldsymbol{\Omega}\right). \tag{90}$$

For a torque-free swimmer, $\boldsymbol{\tau}^{\mathrm{drag}} = 0$, and we obtain a second major result from Rcf. [37]:

$$\boldsymbol{\Omega} = -\frac{3}{8\pi R^4}\int \mathbf{x} \times \mathbf{v}^s(\mathbf{x}_s)\,dS. \tag{91}$$

Finally, we consider how to obtain the stresslet **S**. We multiply Eq. 80 by x_m and integrate over the surface of the sphere:

$$\int u_i^{\infty}(\mathbf{x})x_m\,dS = \int u_i^{\infty}(0)x_m\,dS + \int \left.\frac{\partial u_i^{\infty}}{\partial x_j}\right|_{\mathbf{x}=0} x_m x_j\,dS. \tag{92}$$

The first integral on the right hand vanishes, giving

$$\int u_i^{\infty}(\mathbf{x})x_m\,dS = \frac{4\pi R^4}{3}\left.\frac{\partial u_i^{\infty}}{\partial x_m}\right|_{\mathbf{x}=0}. \tag{93}$$

Swapping the indices i and m, we can also write:

$$\int u_m^{\infty}(\mathbf{x})x_i\,dS = \frac{4\pi R^4}{3}\left.\frac{\partial u_m^{\infty}}{\partial x_i}\right|_{\mathbf{x}=0}. \tag{94}$$

Adding these two equations and dividing by two, we obtain

$$\frac{1}{2}\int \left[u_i^{\infty}(\mathbf{x})x_m + u_m^{\infty}(\mathbf{x})x_i\right]dS = \frac{4\pi R^4}{3}e_{im}^{\infty}(0). \tag{95}$$

Accordingly,

$$e_{im}^{\infty}(0) = -\frac{3}{8\pi R^4}\int \left[v_{s,i}(\mathbf{x})x_m + v_{s,m}(\mathbf{x})x_i\right]dS. \tag{96}$$

Using the Faxén Law in Eq. 77, we obtain:

$$\mathbf{S} = -\frac{5\mu}{2R}\int \left[\mathbf{v}_s(\mathbf{x}_s)\mathbf{x} + \mathbf{x}\mathbf{v}_s(\mathbf{x}_s)\right]dS. \tag{97}$$

This is the major result obtained in Ref. 84 via the Lorentz reciprocal theorem. As before, this manuscript provides a novel alternative derivation and interpretation of Eq. 97 in terms of an effective ambient flow field. (Note that, due to the linearity of the Stokes equation, our approach is easily extended to model active particles in a real ambient flow field.)

Author details

William E. Uspal
Department of Mechanical Engineering, University of Hawai'i at Manoa, Honolulu, HI, United States

*Address all correspondence to: uspal@hawaii.edu

References

[1] Sánchez S, Soler L, Katuri J. Chemically powered micro- and nanomotors. Angewandte Chemie, International Edition. 2015;**54**:1414

[2] Moran JL, Posner JD. Phoretic self-propulsion. Annual Review of Fluid Mechanics. 2016;**49**:511

[3] Aubret A, Ramananarivo S, Palacci J. Eppur si muove, and yet it moves: Patchy (phoretic) swimmers. Current Opinion in Colloid & Interface Science. 2017;**30**:8189

[4] Sundararajan S, Lammert PE, Zudans AW, Crespi VH, Sen A. Catalytic motors for transport of colloidal cargo. Nano Letters. 2008;**8**:1271

[5] Baraban L, Tasinkevych M, Popescu MN, Sánchez S, Dietrich S, Schmidt OG. Transport of cargo by catalytic Janus micro-motors. Soft Matter. 2012;**8**:48

[6] Wu J, Balasubramanian S, Kagan D, Manesh KM, Campuzano S, Wang J. Motion-based DNA detection using catalytic nanomotors. Nature Communications. 2010;**1**:36

[7] Solovev AA, Xi W, Gracias DH, Harazim SM, Deneke C, Sanchez S, et al. Self-propelled nanotools. ACS Nano. 2012;**6**:1751

[8] Aubret A, Youssef M, Sacanna S, Palacci J. Targeted assembly and synchronization of self-spinning microgears. Nature Physics. 2018;**14**: 1114-1118

[9] Cates ME, Tailleur J. Motility-induced phase separation. Annual Review of Condensed Matter Physics. 2015;**6**:219

[10] Wensink HH, Dunkel J, Heidenreich S, Drescher K, Goldstein RE, Lowen H, et al. Meso-scale turbulence in living fluids. Proceedings of the National Academy of Sciences. 2012;**109**:14308

[11] Palacci J, Sacanna S, Steinberg AS, Pine DJ, Chaikin PM. Living crystals of light-activated colloidal surfers. Science. 2013;**339**:936

[12] Buttinoni I, Bialké J, Kümmel F, Löwen H, Bechinger C, Speck T. Dynamical clustering and phase separation in suspensions of self-propelled colloidal particles. Physical Review Letters. 2013;**110**(1):238301

[13] He X, Aizenberg M, Kuksenok O, Zarzar LD, Shastri A, Balazs AC, et al. Synthetic homeostatic materials with chemo-mechano-chemical self-regulation. Nature. 2012;**487**:214

[14] Li J, Shklyaev OE, Li T, Liu W, Shum H, Rozen I, et al. Self-propelled nanomotors autonomously seek and repair cracks. Nano Letters. 2015;**15**: 7077

[15] Uspal WE, Popescu MN, Dietrich S, Tasinkevych M. Rheotaxis of spherical active particles near a planar wall. Soft Matter. 2015;**11**:6613

[16] Bechinger C, Di Leonardo R, Löwen H, Reichhardt C, Volpe G, Volpe G. Active particles in complex and crowded environments. Reviews of Modern Physics. 2016;**88**(1):045006

[17] Paxton WF, Kistler KC, Olmeda CC, Sen A, St SK, Angelo YY, et al. Catalytic nanomotors: Autonomous movement of striped nanorods. Journal of the American Chemical Society. 2004;**126**:13424

[18] Howse JR, Jones RAL, Ryan AJ, Gough T, Vafabakhsh R, Golestanian R. Self-motile colloidal particles: From directed propulsion to random walk. Physical Review Letters. 2007;**99**(1): 048102

[19] Brown A, Poon W. Ionic effects in self-propelled Pt-coated Janus swimmers. Soft Matter. 2014;**10**:4016

[20] Ebbens S, Gregory DA, Dunderdale G, Howse JR, Ibrahim Y, Liverpool TB, et al. Electrokinetic effects in catalytic pt-insulator janus swimmers. EPL. 2014; **106**(1):58003

[21] Brooks AM, Tasinkevych M, Sabrina S, Velegol D, Sen A. Shape-directed rotation of homogeneous micromotors via catalytic self-electrophoresis. Nature Communications. 2019;**10**:495

[22] Derjaguin BV, Yalamov YI, Storozhilova AI. Diffusiophoresis of large aerosol particles. Journal of Colloid and Interface Science. 1966;**22**:117

[23] Anderson JL. Colloid transport by interfacial forces. Annual Review of Fluid Mechanics. 1989;**21**:61

[24] Golestanian R, Liverpool T, Ajdari A. Designing phoretic micro- and nano-swimmers. New Journal of Physics. 2007;**9**(1):126

[25] Moran JL, Posner JD. Electrokinetic locomotion due to reaction-induced charge auto-electrophoresis. Journal of Fluid Mechanics. 2011;**680**:31

[26] de Graaf J, Rempfer G, Holm C. Diffusiophoretic self-propulsion for partially catalytic spherical colloids. IEEE Transactions on Nanobioscience. 2015;**14**:272

[27] Brown AT, Poon WCK, Holm C, de Graaf J. Ionic screening and dissociation are crucial for understanding chemical self-propulsion in polar solvents. Soft Matter. 2017;**13**:1200

[28] Popescu MN, Dietrich S, Oshanin G. Confinement effects on diffusiophoretic self-propellers. The Journal of Chemical Physics. 2009;**130**(1):194702

[29] Crowdy DG. Wall effects on self-diffusiophoretic Janus particles: A theoretical study. Journal of Fluid Mechanics. 2013;**735**:473

[30] Uspal WE, Popescu MN, Dietrich S, Tasinkevych M. Self-propulsion of a catalytically active particle near a planar wall: From reflection to sliding and hovering. Soft Matter. 2015;**11**:434

[31] Das S, Garg A, Campbell AI, Howse JR, Sen A, Velegol D, et al. Boundaries can steer active Janus spheres. Nature Communications. 2015;**6**(1):8999

[32] Simmchen J, Katuri J, Uspal WE, Popescu MN, Tasinkevych M, Sánchez S. Topographical pathways guide chemical microswimmers. Nature Communications. 2016;**7**(1):10598

[33] Liu C, Zhou C, Wang W, Zhang HP. Bimetallic microswimmers speed up in confining channels. Physical Review Letters. 2016;**117**(1):198001

[34] Mozaffari A, Sharifi-Mood N, Koplik J, Maldarelli C. Self-propelled colloidal particle near a planar wall: A Brownian dynamics study. Physical Review Fluids. 2018;**3**:014104

[35] Popescu MN, Uspal WE, Bechinger C, Fischer P. Chemotaxis of active janus nanoparticles. Nano Letters. 2018;**18**: 5345

[36] Palacci J, Sacanna S, Abramian A, Barral J, Hanson K, Grosberg AY, et al. Artificial rheotaxis. Science Advances. 2015;**1**(1):e1400214

[37] Stone HA, Samuel ADT. Propulsion of microorganisms by surface distortions. Physical Review Letters. 1996;**77**:4102

[38] Lighthill MJ. On the squirming motion of nearly spherical deformable bodies through liquids at very small Reynolds numbers. Communications on Pure and Applied Mathematics. 1952;**5**: 109

[39] Blake JR. A spherical envelope approach to ciliary propulsion. Journal of Fluid Mechanics. 1971;**46**:199

[40] Ishikawa T, Locsei JT, Pedley TJ. Development of coherent structures in concentrated suspensions of swimming model micro-organisms. Journal of Fluid Mechanics. 2008;**615**:401

[41] Ishimoto K, Gaffney EA. Squirmer dynamics near a boundary. Physical Review E. 2013;**88**:062702

[42] Katuri J, Uspal WE, Simmchen J, López AM, Sánchez S. Cross-stream migration of active particles. Science Advances. 2018;**4**:eaao1755

[43] Uspal WE. Theory of light-activated catalytic Janus particles. The Journal of Chemical Physics. 2019;**150**:114903

[44] Schaar K, Zöttl A, Stark H. Detention times of microswimmers close to surfaces: Influence of hydrodynamic interactions and noise. Physical Review Letters. 2015;**115**:038101

[45] Mirkovic T, Zacharia NS, Scholes GD, Ozin GA. Nanolocomotion - Catalytic Nanomotors and Nanorotors. Small. 2010;**6**:159

[46] Ebbens SJ, Howse JR. In pursuit of propulsion at the nanoscale. Soft Matter. 2010;**6**:726

[47] Lee T, Alarc'on-Correa M, Miksch C, Hahn K, Gibbs JG, Fischer P. Self-Propelling Nanomotors in the Presence of Strong Brownian Forces. Nano Letters. 2014;**14**:2407-2412

[48] Reigh SY, Huang M, Schofield J, Kapral R. Microscopic and continuum descriptions of Janus motor fluid flow fields. Philosophical Transactions of the Royal Society A. 2016;**374**

[49] de Buyl P, Mikhailov AS, Kapral R. Self-propulsion through symmetry breaking. EPL. 2013;**103**

[50] Dey KK, Das S, Poyton MF, Sengupta S, Butler PJ, Cremer PS, et al. Chemotactic Separation of Enzymes. ACS Nano. 2014;**8**

[51] Pozrikidis C. A Practical Guide to Boundary Element Methods with the Software Library BEMLIB. Boca Raton: CRC Press; 2002

[52] Spagnolie SE, Lauga E. Hydrodynamics of self-propulsion near a boundary: Predictions and accuracy of far-field approximations. Journal of Fluid Mechanics. 2012;**700**:105

[53] Shum H, Gaffney EA. Hydrodynamic analysis of flagellated bacteria swimming in corners of rectangular channels. Physical Review E. 2015;**92**:063016

[54] Montenegro-Johnson TD, Michelin S, Lauga E. A regularised singularity approach to phoretic problems. European Physical Journal E: Soft Matter and Biological Physics. 2015;**38**: 139

[55] Schmiedling L, Lauga E, Montenegro-Johnson TD. Autophoretic flow on a torus. Physical Review Fluids. 2017;**2**:034201

[56] Singh DP,Uspal WE, Popescu MN, Wilson LG and Fischer P. Photo-gravitactic microswimmers. Advanced Functional Materials. 2018;**28**:1706660

[57] Varma A, Montenegro-Johnson TD, Michelin S. Clustering-induced self-propulsion of isotropic autophoretic particles. Soft Matter. 2018;**14**:7155

[58] Popescu MN, Uspal WE, Dietrich S. Self-diffusiophoresis of chemically active colloids. The European Physical Journal Special Topics. 2016;**225**:2189

[59] Kim S, Karrila SJ. Microhydrodynamics: Principles and Selected Applications. Mineola, NY: Dover Publications; 2005

[60] Jackson JD. Classical Electrodynamics. 3rd ed. Hoboken, NJ, USA: Wiley; 1999

[61] Cortez R. The method of regularized Stokeslets. SIAM Journal on Scientific Computing. 2001;**23**:1024

[62] Aderogba K, Blake J. Action of a force near the planar surface between two semi-infinite immiscible liquids at very low Reynolds numbers. Bulletin of the Australian Mathematical Society. 1978;**18**:345

[63] Mathijssen AJTM, Doostmohammadi A, Yeomans JM, Shendruk TN. Hotspots of boundary accumulation: Dynamics and statistics of micro-swimmers in flowing films. Journal of the Royal Society Interface. 2016;**13**:20150936

[64] Delong S, Usabiaga FB, Delgado-Buscalioni R, Griffith BE, Donev A. Brownian dynamics without Green's functions. The Journal of Chemical Physics. 2014;**140**:134110

[65] Rapaport DC. Molecular dynamics simulation using quaternions. Journal of Computational Physics. 1985;**60**:306

[66] Wittkowski R, Löwen H. Self-propelled Brownian spinning top: Dynamics of a biaxial swimmer at low Reynolds numbers. Physical Review E. 2012;**85**:021406

[67] Beard DA, Schlick T. Unbiased rotational moves for rigid-body dynamics. Biophysical Journal. 2003;**85**: 2973

[68] Jones RB, Alavi FN. Rotational diffusion of a tracer colloid particle: IV. Brownian dynamics with wall effects. Physica A. 1992;**187**:436-455

[69] Lisicki M, Cichocki B, Rogers SA, Dhont JKG, Lang PR. Translational and rotational near-wall diffusion of spherical colloids studied by evanescent wave scattering. Soft Matter. 2014;**10**:4312

[70] Fixman M. Implicit algorithm for brownian dynamics of polymers. Macromolecules. 1986;**19**:1195

[71] Bao Y, Rachh M, Keaveny EE, Greengard L, Donev A. A fluctuating boundary integral method for Brownian suspensions. Journal of Computational Physics. 2018;**334**

[72] Uspal WE, Popescu MN, Tasinkevych M, Dietrich S. Shape-dependent guidance of active Janus particles by chemically patterned surfaces. New Journal of Physics. 2018; **20**:015013

[73] Brady JS, Bossis G. Stokesian dynamics. Annual Review of Fluid Mechanics. 1988;**20**:111

[74] Kim AS, Stolzenbach KD. The Permeability of Synthetic Fractal Aggregates with Realistic Three-Dimensional Structure. Journal of Colloid and Interface Science. 2002;**253**: 315328

[75] Kim AS, Stolzenbach KD. Aggregate formation and collision efficiency in differential settling. Journal of Colloid and Interface Science. 2004;**271**:110

[76] Yan W, Brady JF. The force on a boundary in active matter. Journal of Fluid Mechanics. 2015;**785**:1

[77] Walter J, Salsac A-V, Barthés-Biesel D, Le Tallec P. Coupling of finite element and boundary integral methods for a capsule in a Stokes flow. International Journal for Numerical Methods in Engineering. 2010;**83**:829

[78] Gao T, Hu HH. Deformation of elastic particles in viscous shear flow.

Journal of Computational Physics. 2009; **228**:2132

[79] Zhao H, Isfahani AHG, Olson LN, Freund JB. A spectral boundary integral method for flowing blood cells. Journal of Computational Physics. 2010;**229**: 3726

[80] Würger ACR. Is Soret equilibrium a non-equilibrium effect?. Mécanique. 2013;**341**:438

[81] Nikola N, Solon AP, Kafri Y, Kardar M, Tailleur J, Voituriez R. Active particles with soft and curved walls: equation of state, ratchets, and instabilities. Physical Review Letters. 2016;**117**:098001

[82] Montenegro-Johnson TD. Microtransformers: Controlled microscale navigation with flexible robots. Physical Review Fluids. 2018;**3**: 062201(R)

[83] Batchelor GK, Green JT. The hydrodynamic interaction of two small freely-moving spheres in a linear flow field. Journal of Fluid Mechanics. 1972; **56**:375

[84] Lauga E, Michelin S. Stresslets induced by active swimmers. Physical Review Letters. 2016;**117**:148001

Chapter 3

Fundamentals of Irreversible Thermodynamics for Coupled Transport

Albert S. Kim

Abstract

Engineering phenomena occur in open systems undergoing irreversible, non-equilibrium processes for coupled mass, energy, and momentum transport. The momentum transport often becomes a primary or background process, on which driving forces of physical gradients govern mass and heat transfer rates. Although in the steady state no physical variables have explicit variation with time, entropy increases with time as long as the systems are open. The degree of irreversibility can be measured by the entropy-increasing rate, first proposed by L. Onsager. This book conceptually reorganizes the entropy and its rate in broader aspects. Diffusion is fully described as an irreversible, i.e., entropy increasing, phenomenon using four different physical pictures. Finally, an irreversible thermodynamic formalism using effective driving forces is established as an extension to the Onsager's reciprocal theorem, which was applied to core engineering phenomena of fundamental importance: solute diffusion and thermal flux. In addition, the osmotic and thermal fluxes are explained in the unified theoretical framework.

Keywords: irreversible thermodynamics, non-equilibrium thermodynamics, Onsager's reciprocal theorem, entropy rate, diffusion pictures, irreversible transport equation

1. Introduction

This chapter contributes to a comprehensive explanation of the steady-state thermodynamics of irreversible processes with detailed theoretical derivations and examples. The origin and definitions of entropy are described, irreversible thermodynamics for a steady state is revisited based on Onsager's reciprocal theorem, and thermal and solute diffusion phenomena are recapitulated as examples of single-component irreversible thermodynamic processes.

1.1 Thermodynamic states

In fundamental and applied sciences, thermodynamics (or statistical mechanics) plays an important role in understanding macroscopic behaviors of a thermodynamic system using microscopic properties of the system. Thermodynamic systems have three classifications based on their respective transport conditions at interfaces.

An *open* system allows energy and mass transfer across its interface, a *closed* system allows transfer of energy only, but preventing mass transfer, and, finally, an *isolated* system allows no transport across its interface.

Transfer phenomenon of mass and energy are represented using the concept of flux, which is defined as a rate of passing a physical variable of interest across a unit cross-sectional area per unit time. If the flux is constant, input and output rates of a physical quantity within a finite volume are equal, and the density remains constant because a net accumulation within the systems is zero. If the flux varies spatially, specifically $\boldsymbol{J} = \boldsymbol{J}(x, y, z)$, then its density within the specified volume changes with time, i.e., $\rho = \rho(t)$. This balance is defined as the equation of continuity:

$$\frac{\partial \rho}{\partial t} + \nabla \cdot \boldsymbol{J} = 0 \tag{1}$$

Many engineering processes occur in an open environment, having specific mass and energy transfer phenomena as practical goals. An exception is a batch reaction, where interfacial transport is blocked and a transient variation of the internal system is of concern. If the internal characteristic of the open system changes with time, the system moves toward a transient, non-equilibrium state. However, the transiency is subject to the human perception of the respective time scale. If engineering system performance is averaged over a macroscopic time scale, such as hours, days, and weeks, the time-averaged performance is a primary concern as those quantities can be compared with experimental data. Instead of transiency, the time to reach a steady state becomes more important in operating engineering processes because a steady-state operation is usually sought. Usually, the time to reach a steady state is much shorter than the standard operation time in a steady state.

1.2 Time scale and transiency

In theoretical physics, statistical mechanics and fluid dynamics are not fully unified, and non-equilibrium thermodynamics is unsolved in theoretical physics. It is often assumed that the fluid flow is not highly turbulent, and a steady state is reached with a fully developed flow field. The thermodynamic characteristics are maintained within the steady flow, and the static equilibrium is assumed to be valid within small moving fluid elements. In such a situation, each fluid element can be qualitatively analogous to a microstate of the thermodynamic ensemble.

Nevertheless, a conflict between the thermodynamics and fluid dynamics stems from the absence of a clear boundary between the static equilibrium for isolated systems and the steady state of open systems. In principle, the steady state belongs to the non-equilibrium state although the partial differentials of any physical quantities are assumed to be zero (i.e., $\partial/\partial t = 0$). A density does not change with time, but the flux exists as finite and constant in time and space. The time scale of particle motion can be expressed using the particle relaxation time defined as $\tau_p = m/\beta$, where m and β exist as particle mass and Stokes' drag coefficient, respectively. The time scale for the fluid flow can be evaluated as the characteristic length divided by the mean flow speed, but the particle relaxation time scale is much shorter than the flow time scale. Therefore, the local equilibrium may be applied without significant deviation from the real thermodynamic state.

In engineering, various dimensionless numbers are often used to characterize a system of interest. The Reynolds (Re) and Péclet (Pe) numbers indicate ratios of the convective transport to viscous momentum and diffusive heat/mass transport in a fluid, respectively. The Nusselt and Sherwood numbers represent ratios of the diffusion length scale as compared to the boundary layer thickness of the thermal and mass diffusion phenomenon, respectively. The Prandtl and Schmidt numbers represent ratios of momentum as compared to thermal and mass diffusivities, respectively. Other dimensionless numbers include the Biot number (Bi) (for both heat and mass transfer), the Knudsen number (Kn) (molecular mean free path to system length scale), the Grashof (Gr) number (natural buoyancy to viscous forces), and the Rayleigh number (natural convective to diffusive heat/mass transfer).

Note that all the dimensionless numbers described here implicitly assume the presence of fluid flow in open systems because they quantify the relative significance of energy, momentum, and mass transport. The static equilibrium approximation (SEA) must be appropriate if the viscous force is dominant within a fluid region, preventing transient system fluctuation, as the non-equilibrium thermodynamics is not fully established in theoretical physics and steady-state thermodynamics requires experimental observations to determine thermodynamic coefficients between driving forces and generated fluxes.

1.3 Statistical ensembles

Thermodynamics often deals with macroscopic, measurable phenomena of systems of interest, consisting of objects (e.g., molecules or particles) within a volume. Statistical mechanics is considered as a probabilistic approach to study the microscopic aspects of thermodynamic systems using microstates and ensembles and to explain the macroscopic behavior of the respective systems.

Seven variables exist within statistical mechanics (i.e., temperature T, pressure P, and particle number N, which are conjugated to entropy S, volume V, chemical potential μ, and finally energy E of various forms). The thermodynamic ensemble uses the first and second laws of thermodynamics and provides constraints of having three out of the six variables (excluding E) remaining constant. The other three conjugate variables are theoretically calculated or experimentally measured. Statistical ensembles are either isothermal (for constant temperature) or adiabatic (of zero heat exchanged at interfaces). The adiabatic category includes NVE (microcanonical), NPH, μVL, and μPR ensembles, and isothermal ensembles possess NVT (canonical), NPT (isobaric-isothermal or Gibbs), and μVT (grand canonical). Here, ensembles of NVE and NPH are called microcanonical and isenthalpic, and those of NVT, μVT, and NPT are called canonical, grand canonical, and isothermal-isobaric, respectively. Within statistical mechanical theories and simulations, canonical ensembles are most widely used, followed by grand canonical and isothermal-isobaric ensembles. The adiabatic ensembles are equivalent to isentropic ensembles (of constant entropy) and are represented as NVS, NPS, μVS and μPS instead of NVE, NPH, μVL, and μPR, respectively. Non-isothermal ensembles often represent entropy S as a function of a specific energy form, of which details can be found elsewhere [1].

2. Entropy revisited

2.1 Thermodynamic laws

Thermodynamic laws can be summarized as follows:

- The zeroth law: For thermodynamic systems of A, B, and C, if $A = C$ and $B = C$, then $A = B$.

- The first law: The internal energy change ΔU is equal to the energy added to the system Q, subtracted by work done by the system W (i.e., $\Delta U = Q - W$).

- The second law: An element of irreversible heat transferred, δQ, is a product of the temperature T and the increment of its conjugate variable S (i.e., $\delta Q = TdS$).
- The third law: As $T \to 0$, $S \to$ constant, and $S = k_B \ln \Omega$, where Ω is the number of microstates.

The entropy S is defined in the second thermodynamic laws, and its fundamental property is described in the third law, linking the macroscopic element of irreversible heat transferred (i.e., δQ) and the microstates of the system.

Suppose you have N objects (e.g., people) and need to position them in a straight line consisting of the same number of seats. The first and second objects have N and $N-1$ choices, respectively; similarly, the third one has $N-2$; the fourth one has $N-3$ choices; and so on. The total number of ways of this experiment is as follows:

$$N \cdot (N-1) \cdot (N-2) \cdot (N-3) \cdot \cdots \cdot 2 \cdot 1 = N! \tag{2}$$

Example 1: In a car, there are four seats including a driver's. Three guests will occupy the same number of seats. How many different configurations are available? There are three people, A, B, and C, and three seats, S_1, S_2, and S_3. If A can chose a seat first, then A has three choices. Then, B and C have, in a sequence, two and one choices. Then, the total number of possible configurations are $3 \cdot 2 \cdot 1 = 3! = 6$.

Next, when the N objects are divided into two groups. Group 1 and group 2 can contain N_1 and N_2 objects, respectively. Then, the total number of the possible ways to place N objects into two groups is

$$\frac{N!}{N_1!N_2!}$$

which is equal to the number of combinations of N objects taking N_1objects at a time

$$C_{N_1}^{N} = \frac{N!}{N_1!(N-N_1)!} \tag{3}$$

For example, consider the following equation of a binomial expansion

$$(x+y)^3 = 1 \cdot x^3 + 3x^2y + 3xy^2 + 1 \cdot y^3 = \sum_{n=0}^{3} a_n x^n y^{3-n} \tag{4}$$

where $a_0 = a_3 = 1$ and $a_1 = a_2 = 3$. For the power of N, the equation exists as

$$(x_1 + x_2)^N = \sum_{N_1=0} \sum_{N_2=0} \frac{N!}{N_1!N_2!} x_1^{N_1} x_2^{N_2} = \sum_{k=0}^{N} C_N^k x_1^k x_2^{N-k} \tag{5}$$

where $N_1 + N_2 = N$ and

$$C_N^k = \frac{N!}{k!(N-k)!} = C_N^{N-k} \tag{6}$$

If we add x_3 with a constraint condition of $N = \sum_{k=1}^{3} N_k$, then

$$(x_1 + x_2 + x_3)^N = \sum_{N_1=0} \sum_{N_2=0} \sum_{N_3=0} \frac{N!}{N_1!N_2!N_3!} x_1^{N_1} x_2^{N_2} x_3^{N_3} \tag{7}$$

where the coefficient of the polynomial expansion can be written as follows:

$$\frac{N!}{N_1!N_2!N_3!} = \frac{N!}{\prod_{k=1}^{3} N_k!} \tag{8}$$

using the product notation of

$$y_1 \cdot y_2 \cdot y_3 \cdot \cdots \cdot y_m = \prod_{k=1}^{m} y_k \tag{9}$$

Example 2: Imagine that we have three containers and ten balls. Each container has enough room to hold all ten balls. Let N_i (for $i = 1 - 3$) be the number of balls in i^{th} container. How many different configurations are available to put ten balls into the three containers? If $N_1 = 2$, $N_2 = 3$, and $N_3 = 5$, then the equation is with the answer being 2520:

$$\frac{N!}{N_1!N_2!N_3!} = \frac{10!}{2!3!5!} = \frac{3628800}{(2)(6)(120)} = 2520 \tag{10}$$

satisfying $N = N_1 + N_2 + N_3 = 10$.

2.2 Definitions

2.2.1 Boltzmann's entropy

A thermodynamic system is assumed to have a number of small micro-systems. Say that there are N micro-systems and $m\,(\leq N)$ thermodynamic states. This situation is similar to $N(=10)$ balls in $m(=3)$ containers. The number of balls in container 1, 2, and 3 is N_1, N_2, and N_3, respectively. Then the total number of different configurations of micro-systems in m micro-states is defined as

$$\Omega_N = \frac{N!}{\prod_{k=1}^{m} N_k!} \tag{11}$$

Boltzmann proposed a representation of entropy of the entire ensemble as

$$S_B = k_B \ln \Omega_N \tag{12}$$

2.2.2 Gibbs entropy

The Gibbs entropy can be written using Ω, as

$$\frac{S}{k_B} = \ln \Omega_N = \ln \frac{N!}{\prod_{k=0}^{m} N_k!} = \ln N! - \sum_{k=0}^{m} \ln N_k!$$

and using Stirling's formula as

$$\ln N! = N \ln N/e$$

for a large $N\,(\gg 1)$, to derive

$$\frac{S}{k_B} \simeq N \ln (N/e) - \sum_{k=0}^{m} N_k \ln (N/e) = -N \sum_{k=0}^{m} \left(\frac{N_k}{N}\right) \cdot \ln \left(\frac{N_k}{N}\right)$$

Finally, we have

$$S = -k_B N \sum_{k=0}^{m} p_k \ln p_k$$

where $p_k = N_k/N$ exists as the probability of finding the system in thermodynamic state k. Gibbs introduced a form of entropy as

$$s_G = -k_B \sum_{k=0}^{m} p_k \ln p_k$$

which is equal to the system entropy per object or particle, denoted as

$$s_G = \frac{S}{N} = -k_B \sum_{k=0}^{m} p_k \ln p_k$$

2.2.3 Shannon's entropy

In information theory, Shannon's entropy is defined as [2]

$$S_{\mathrm{Sh}} = -\sum_i p_i \log_b p_i \qquad (13)$$

As the digital representation of integers is binary, the base b is often set as two. Note that Shannon's entropy is identical to Gibbs entropy, if Boltzmann's constant k_B is discarded and the natural logarithm $\ln = \log_e$ is replaced by $\log_2$. Entropy only takes into account the probability of observing a specific event, so the information it encapsulates is information about the underlying probability distribution, not the meaning of the events themselves. Example 3 deals with tossing a coin or a dice and how the entropy S increases with respect to the number of available outcomes.

Example 3: Let's consider two conventional examples, i.e., a coin and a dice. Their Gibbs entropy values (i.e., entropy per object) are

$$\frac{s_{\mathrm{coin}}}{k_B} = -\sum_{k=1}^{2} p_k \ln p_k = -\sum_{k=1}^{2} \left(\frac{1}{2} \cdot \ln \frac{1}{2}\right) = \ln 2 = 0.6931$$

$$\frac{s_{\mathrm{dice}}}{k_B} = -\sum_{k=1}^{6} \cdot \frac{1}{6} \ln \frac{1}{6} = \ln 6 = 1.791$$

The system entropies of the coin and the dice are

$$S_{\mathrm{coin}}/k_B = 2 \times 0.6931 = 1.386$$

$$S_{\mathrm{dice}}/k_B = 6 \times 1.791 = 10.750$$

and their ratio is

$$\frac{S_{\mathrm{dice}}}{S_{\mathrm{coin}}} = \frac{6 \times \ln 6}{2 \times \ln 2} = \mathbf{3} \cdot \frac{\ln 2 \cdot \mathbf{3}}{\ln 2} = \mathbf{3} \times 2.5850 = 7.754 > \mathbf{3}$$

where three indicates the ratio of the number of available cases of a dice (6) to that of a coin (2). The entropy ratio, 7.754, is higher than the ratio of available states, 3.

3. Diffusion: an irreversible phenomenon

Diffusion refers to a net flow of matter from a region of high concentration to a region of low concentration, which is an entropy-increasing process, from a more ordered to a less ordered state of molecular locations. For example, when a lump of sugar is added to a cup of coffee for a sweeter taste, the solid cube of sugar dissolves, and the molecules spread out until evenly distributed. This change from a localized to a more even distribution exists as a spontaneous and, more importantly, irreversible process. In other words, diffusion occurs by itself without external driving forces. In addition, once diffusion occurs, it is not possible for the molecular distribution to return to its original undiffused state. If diffusion does not occur spontaneously, then there is no natural mixing, and one may have a bitter coffee taste and sweet sugar taste in an unmixed liquid phase. In general, diffusion is closely related to mixing and demixing (separation) within a plethora of engineering applications. Why does diffusion occur, and how do we understand the spontaneous phenomena? A key stands as the entropy-changing rate from one static equilibrium to the other. Before discussing diffusion as an irreversible phenomenon, however, the following section includes several pictures so as to create a better understanding of diffusion phenomenon as one of the irreversible thermodynamic processes.

3.1 Mutual diffusion

Diffusion is often driven by the concentration gradient referred to as ∇c, typically in a finite volume, temperature, and pressure. As temperature increases, molecules gain kinetic energy and diffuse more actively in order to position evenly within the volume. A general driving force for isothermal diffusion exists as a gradient of the chemical potential $\nabla \mu$ between regions of higher and lower concentrations.

As shown in **Figure** 1, diffusion of solute molecules after removing the mid-wall is spontaneous. Initially, two equal-sized rectangular chambers A and B are separated by an impermeable wall between them. The thickness of the mid-wall is negligible in comparison to the box length; in each chamber of A and B, the same amount of water is contained. Chamber A contains seawater of salt concentration 35,000 ppm, and chamber B contains fresh water of zero salt concentration. If the separating wall is removed slowly enough not to disturb the stationary solvent medium but fast enough to initialize a sharp concentration boundary between the

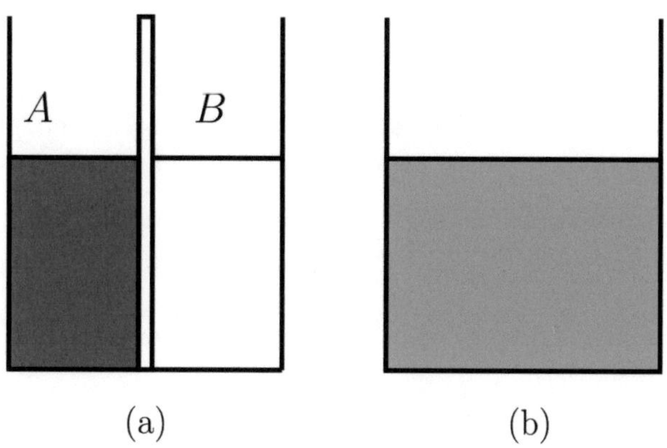

Figure 1.
Diffusion in a rectangular container consisting of two equal-sized chambers A and B (a) before and (b) after the mid-wall is removed.

two concentration regions, then the concentration in B increases as much as that in A decreases because mass is neither created nor annihilated inside the container. This spontaneous mixing continues until both concentrations become equal and, hence, reach a thermodynamic equilibrium consisting of a half-seawater/half-fresh water concentration throughout the entire box. Diffusion occurs wherever and whenever the concentration gradient exists, and diffusive solute flux is represented using Fick's law as follows [3, 4]:

$$J_s = -D\frac{dc}{dx} \quad \text{in } 1-D \tag{14}$$

or

$$\mathbf{J}_s = -D\nabla c \quad \text{in } 3-D \tag{15}$$

where D is diffusion coefficient (also often called diffusivity) of a unit of m^2/s. A length scale of diffusion can be estimated by $\sqrt{D\delta t}$ where δt is a representative time interval. In molecular motion, δt can be interpreted as a time duration required for a molecule to move as much as a mean free path (i.e., a statistical averaged distance between two consecutive collisions).

3.2 Stokes-Einstein diffusivity

When the solute concentration is low so that interactions between solutes are negligible, the diffusion coefficient, known as the Stokes-Einstein diffusivity, may be given by

$$D_0 = \frac{k_B T}{6\pi\eta a} \tag{16}$$

where k_B is the Boltzmann constant, η is the solvent viscosity[1], and a is the (hydrodynamic) radius of solute particles. Stokes derived hydrodynamic force that a stationary sphere experiences when it is positioned in an ambient flow [5]:

$$F_H = 6\pi\eta a v \tag{17}$$

where v represents a uniform fluid velocity, which can be interpreted as the velocity of a particle relative to that of an ambient fluid. F_H is linearly proportional to v, and its proportionality $6\pi\eta a$ is the denominator of the right-hand side of Eq. (16). Einstein used the transition probability of molecules from one site to the another, and Langevin considered the molecular collisions as random forces acting on a solute (see Section 3.3 for details). Einstein and Langevin independently derived the same equation as (16) of which the general form can be rewritten as

$$D_0 = \frac{k_B T}{(2d)\pi\eta} \tag{18}$$

where d is the spatial dimension of the diffusive system (i.e., $d = 1, 2,$ and 3 for 1D, 2D, and 3D spaces).

[1] Greek symbol μ is also often used for viscosity in fluid mechanics literature. In this book, chemical potential is denoted as μ.

3.3 Diffusion pictures

Several pictures of diffusion phenomena are discussed in the following section, which give probabilistic and deterministic viewpoints. If one considers an ideal situation where there exists only one salt molecule in a box filled with solvent (e.g., water) of finite T, P, and V. Since the sole molecule exists, there is no concentration gradient. Mathematically, the concentration is infinite at the location of the molecule and absolutely zero anywhere else: $c = V^{-1}\delta(\boldsymbol{r} - \boldsymbol{r}_0)$ where $\boldsymbol{r}_0$ is an initial position of the solute and $\boldsymbol{r}$ is an arbitrary location within the volume. However, the following question arises. Why does a single molecule diffuse without experiencing any collisions in the absence of other molecules? The answer is that the solvent medium consists of a number of (water) molecules having a size of an order of $O(10^{-10})$ m. The salt molecule will suffer a tremendous number of collisions with solvent molecules of a certain kinetic energy at temperature T. Since each of these collisions can be thought of as producing a jump of the molecule, the molecule must be found at a distance from its initial position where the diffusion started. In this case, the molecule undergoes Brownian motion. Note that the single molecule collides only with solvent molecules while diffusing, which exists as a type of diffusion called self-diffusion.

3.3.1 Self-diffusion and random walk

A particle initially located at $\boldsymbol{r}_0$ has equal probabilities of 1/6 to move in $(\pm x, \pm y, \pm z)$ directions. For mathematical simplicity, we restrict ourselves to 1D random walk of a dizzy individual, who moves to the right or to the left with a 50:50 chance. Initially (at time $t = 0$), the individual is located at $x_0 = 0$ and starts moving in a direction represented by $\Delta x = \pm l$ where $+l$ and $-l$ indicate the right and left distances that the individual travels with an equal probability, respectively. At the next step, $t_1 = t_0 + \Delta t = \Delta t$, the individual's location is found at

$$x_1 = x_0 + \Delta x_1 = \Delta x_1 \tag{19}$$

where Δx_1 can be $+l$ or $-l$. At the time of the second step, $t_2 = t_1 + \Delta t = 2\Delta t$, the position is

$$x_2 = x_1 + \Delta x_2 = \Delta x_1 + \Delta x_2 \tag{20}$$

where $\Delta x_2 = \pm l$. At $t_n = n\Delta t$ $(n \gg 1)$, the position may be expressed as

$$x_n = \Delta x_1 + \Delta x_2 + \cdots + \Delta x_{n-1} + \Delta x_n = \sum_{i=1}^{n} \Delta x_i \tag{21}$$

If there are a number of dizzy individuals and we can determine an average for their seemingly random movements, then

$$\langle x_n \rangle = \sum_{i=1}^{n} \langle \Delta x_i \rangle = n\langle \Delta x \rangle = 0 \tag{22}$$

because Δx has a 50:50 chance of $+l$ and $-l$:

$$\langle \Delta x \rangle = (+l)\frac{1}{2} + (-l)\frac{1}{2} = 0 \tag{23}$$

Now let us calculate a mean of x^2:

$$\begin{aligned}\langle x_n^2\rangle &= \langle(\Delta x_1 + \cdots \Delta x_n)\cdot(\Delta x_1 + \cdots \Delta x_n)\rangle \\ &= \langle \Delta x_1^2 + \Delta x_1\cdot\Delta x_2 + \cdots + \Delta x_1\cdot\Delta x_n \\ &\quad + \Delta x_2\cdot\Delta x_1 + \Delta x_2^2 + \cdots + \Delta x_2\cdot\Delta x_n \\ &\quad + \cdots \\ &\quad + \Delta x_n\cdot\Delta x_1 + \Delta x_n\cdot\Delta x_2 + \cdots + \Delta x_n^2\rangle\end{aligned} \tag{24}$$

and in a concise form

$$\langle x_n^2\rangle = \left\langle \sum_{i\neq j}\Delta x_i\cdot\Delta x_j\right\rangle + \left\langle \sum_{k=1}^{n}\Delta x_k^2\right\rangle = 0 + n\Delta x^2 = nl^2 \tag{25}$$

because $\left\langle \sum_{i\neq j}\Delta x_i\cdot\Delta x_j\right\rangle = 0$ and $\langle \Delta x_k^2\rangle = (\Delta x)^2 = l^2$. In the calculation of off-diagonal terms, $\Delta x_i\cdot\Delta x_j$ can have four possible values with equal chance of $(+,+)$, $(+,-)$, $(-,+)$, and $(-,-)$. The products of the two elements in the parenthesis are $+$, $-$, $-$, and $+$ with equal probability of 25%. Therefore, a sum of them is zero. Because n is the number of time steps, it can be replaced by $t/\Delta t$ where t is the total elapsed time. The diffusion coefficient in one-dimensional space was derived in the previous section as $D = l^2/2\Delta t$. Then, the mean of squared distance at time t is calculated as

$$\langle x^2(t)\rangle = 2Dt \tag{26}$$

and the root-mean-square distance is

$$x_{\mathrm{rms}} = \sqrt{\langle x^2(t)\rangle} = \sqrt{2Dt} \tag{27}$$

Note that x_{rms} is proportional to $t^{1/2}$ in the random walk, as compared to the constant velocity case $x = vt\propto t^1$. Then, the diffusivity for 1D is explicitly

$$D = \frac{x_{\mathrm{rms}}^2}{2t} = \frac{nl^2}{2n\Delta t} = \frac{l^2}{2\Delta t} \tag{28}$$

3.3.2 Einstein's picture

The concentration $C(x,t)$ after an infinitesimal time duration δt from t within a range dx between x and $x + \mathrm{d}x$ is calculated as [6]

$$C(x,t+\delta t)\mathrm{d}x = \mathrm{d}x\int_{-\infty}^{+\infty} C(x-y,t)\Phi(y)\mathrm{d}y \tag{29}$$

where Φ is the transition probability for a linear displacement y and the right-hand side indicates the amount of adjacent solutes that move into the small region dx. The probability distribution satisfies

$$\int_{-\infty}^{+\infty}\Phi(y)\mathrm{d}y = 1 \tag{30}$$

and we assume that Φ is a short ranged, even function, meaning that it is non-zero for small $|y|$ and symmetric, $\Phi(-y) = \Phi(y)$. In this case, we approximate the integrand of Eq. (29) as

$$C(x-y,t) = C(x,t) - \frac{\partial C}{\partial x}y + \frac{1}{2!}\frac{\partial^2 C}{\partial x^2}y^2 + \cdots \tag{31}$$

and substitute Eq. (31) with Eq. (29). We finally derive the so-called diffusion equation:

$$\frac{\partial C}{\partial t} = D_B \frac{\partial^2 C}{\partial x^2} \tag{32}$$

where the diffusivity is defined as

$$D_B = \frac{\langle y^2 \rangle}{2!\delta t} \tag{33}$$

where $\langle y^2 \rangle$ is the mean value of y^2, calculated as

$$\langle y^2 \rangle = \int_{-\infty}^{+\infty} y^2 \Phi(y) \mathrm{d}y \tag{34}$$

Within this calculation, we used

$$C(x, t+\delta t) = C(x,t) + \frac{\partial C}{\partial t}\delta t + \cdots \tag{35}$$

and

$$\langle y \rangle = \int_{-\infty}^{+\infty} y \Phi(y) \mathrm{d}y = 0 \tag{36}$$

because $y\Phi$ is an odd function. Mathematically, Einstein's picture uses short-ranged transition probability function, which does not need to be specifically known, and Taylor's expansion for a small time interval and short displacement. Conditions required for Eq. (32) are as follows: (i) transition distance is longer than the size of molecule, $\mathrm{d}x \geq O(a)$, and (ii) time interval δt is long enough to measure $\mathrm{d}x$ after a tremendous number of collisions with solvent molecules, satisfying $\delta t \gg \tau_p$, where τ_p is the particle relaxation time (see Langevin's picture).

3.3.3 Langevin's picture

Let us consider a particle of mass m, located at $x(t)$ with velocity $v \equiv \mathrm{d}x/\mathrm{d}t$ at time t. For simplicity, we shall treat the problem of diffusion in one dimension. It would be hopeless to deterministically trace all the collisions of this particle with a number of solvent molecules in series. However, these collisions can be regarded as a net force $A(t)$ effective in determining the time dependence of the molecule's position $x(t)$. Newton's second law of motion can be written in the following form [7, 8]:

$$m\frac{\mathrm{d}v}{\mathrm{d}t} = -\beta v + A(t) \tag{37}$$

which is called Langevin's equation. In Eq. (37), $A(t)$ is assumed to be randomly and rapidly fluctuating. We multiply x on both sides of Eq. (37) to give

$$m x \frac{dv}{dt} = -\beta x v + x A(t) \tag{38}$$

and take a time average of both sides during an interval τ, defined as

$$\langle \cdots \rangle = \frac{1}{\tau} \int_t^{t+\tau} (\cdots) dt \tag{39}$$

Then, we have after a much longer time than the particle relaxation time τ:

$$m \left\langle x \frac{dv}{dt} \right\rangle = -\beta \langle xv \rangle + \langle x A(t) \rangle \tag{40}$$

Because the random fluctuating force $A(t)$ is independent of the particle position $x(t)$, we calculate

$$\langle xA \rangle = \langle x \rangle \langle A \rangle = \langle x \rangle \cdot 0 = 0 \tag{41}$$

For further derivation, we use the following identities:

$$\frac{dx^2}{dt} = 2x\dot{x} = 2xv \tag{42}$$

$$\frac{d^2x^2}{dt^2} = \frac{d}{dt}(2x\dot{x}) = 2v^2 + 2xv \tag{43}$$

to provide

$$m \left\langle \frac{1}{2} \frac{d^2x^2}{dt^2} - v^2 \right\rangle = -\beta \left\langle \frac{1}{2} \frac{dx^2}{dt} \right\rangle \tag{44}$$

We let $z = \langle dx^2/dt \rangle$ and rewrite Eq. (44):

$$m \frac{dz}{dt} = -\beta \left(z - \frac{2k_B T}{\beta} \right) \tag{45}$$

because the kinetic energy of this particle is equal to the thermal energy:

$$\frac{1}{2} m v^2 = \frac{1}{2} k_B T \tag{46}$$

where k_B is the Boltzmann constant. Note that the origin of the particle motion exists as the number of its collisions with solvent molecules at temperature T. If we take an initial condition of $z = 0$ indicating either position or velocity is initially zero, then we obtain

$$z(t) = \frac{2k_B T}{\beta} \left(1 - e^{-t/\tau_p} \right) = \frac{d\langle x^2 \rangle}{dt} \tag{47}$$

where $\tau_p = m/\beta$ is the particle relaxation time. One more integration with respect to time yields

$$\begin{aligned}\langle x^2\rangle &= \frac{2k_BT}{\beta}\int_0^t \left(1-e^{-t'/\tau_p}\right)\mathrm{d}t' \\ &= \frac{2k_BT\tau_p}{\beta}\left[\frac{t}{\tau_p}+e^{-t/\tau_p}-1\right]\end{aligned} \tag{48}$$

If $t \gg \tau_p$, then t/τ_p in the rectangular parenthesis is dominant:

$$\langle x^2\rangle = \frac{2k_BT}{\beta}t \equiv 2D_Bt \tag{49}$$

Stokes' law of Eq. (17) indicates $\beta = 6\pi\eta a$, and, therefore, the diffusion coefficient of Brownian motion or Stokes-Einstein diffusivity is

$$D_B = \frac{k_BT}{6\pi\eta a} \tag{50}$$

identical to Eq. (16). The root-mean-square distance is

$$x_{\mathrm{rms}} = \sqrt{\langle x^2\rangle} = \sqrt{2D_Bt} \tag{51}$$

which is proportional to $\sqrt{t}$. Note that $\langle x\rangle = 0$. From an arbitrary time t, the particle drifts for an interval Δt, where $\Delta t \gg \tau_p$, and then

$$x_{\mathrm{rms}}(\Delta t) = \sqrt{\langle x^2(t+\Delta t)\rangle - \langle x^2(t)\rangle} = \sqrt{2D_B\Delta t} \tag{52}$$

Then, the time step Δt is of a macroscopic scale in that one can appreciate the movement of the particle of an order of particle radius. For a short time $t \ll \tau_p$, the mean-square distance of Eq. (48) is approximated as $x_{\mathrm{rms}} = v_{\mathrm{rms}}t$, indicating a constant velocity motion.

Einstein's and Langevin's pictures provide identical results for x_{rms} and D_B as related to Stokes' law. On one hand, if a particle is translating with a constant velocity, its distance from the initial location is linearly proportional to the elapsed time; on the other hand, if particle is diffusing, its root-mean-square distance is proportional to $\sqrt{t}$.

3.3.4 Gardiner's picture

In Langevin's Eq. (37), the randomly fluctuating force can be written as

$$A(t) = \alpha f(t) \tag{53}$$

where f satisfies

$$\langle f(t)\rangle = \frac{1}{T_p}\int_0^{T_p} f(t)\mathrm{d}t = 0 \quad \text{for} \quad T_p \gg \tau_p \tag{54}$$

and

$$\left\langle f_i(t) f_j(t)\right\rangle = \delta_{ij}\delta(t-t') \tag{55}$$

Relationships between parameters are

$$\beta = 6\pi\mu a_p \tag{56}$$

$$\alpha = \sqrt{2\beta k_B T} = \sqrt{2D_B^{-1} k_B T} \tag{57}$$

$$D_B = \frac{k_B T}{\beta} \tag{58}$$

(See the next section for the Brownian diffusivity D_B.) As such, we assume that

$$f(t)\mathrm{d}t = \mathrm{d}W(t) \tag{59}$$

where dW is the Ito-Wiener process [9, 10], satisfying

$$\langle \mathrm{d}W \rangle = 0 \tag{60}$$

$$\langle (\mathrm{d}W) \rangle^2 = \mathrm{d}t \tag{61}$$

Then, we can obtain the stochastic differential equation (SDE) as

$$m\mathrm{d}v = [F(x) - \beta v]\mathrm{d}t + \alpha \mathrm{d}W \tag{62}$$

The relationship between x, v, and t can be obtained as follows [11]:

$$\mathrm{d}x = v\mathrm{d}t \tag{63}$$

$$\mathrm{d}v = \left[\frac{F(x)}{m} - \frac{v}{\tau_p}\right]\mathrm{d}t + \frac{\alpha}{m}\mathrm{d}W \tag{64}$$

Note that Eq. (63) is free from the fundamental restriction of Langevin's equation (i.e., $\tau_p \ll \mathrm{d}t$) by introducing the Ito-Wiener process in Eq. (64). The time interval dt can be arbitrarily chosen to improve calculation speed and/or numerical accuracy.

Eq. (63) uses the basic definition of velocity as a time derivative of the position in the classical mechanics, and Eq. (64) represents the randomly fluctuating force using the Ito-Weiner process, dW. If we keep Langevin's picture, then these two equations should have forms of

$$\mathrm{d}x = v\mathrm{d}t + \sqrt{2D_B}\mathrm{d}W \tag{65}$$

$$\mathrm{d}v = \left[\frac{F(x)}{m} - \frac{v}{\tau_p}\right]\mathrm{d}t \tag{66}$$

where the random fluctuation disappears in the force balance and appears as a drift displacement, $\sqrt{2D_B}\mathrm{d}W$. Let $C(x)$ be the concentration of particles near the position x of a specific particle. Note that x is not a fixed point in Eulerian space but a moving coordinate of a particle being tracked. An infinitesimal change of C is

$$\mathrm{d}C(x) = C'\mathrm{d}x + \frac{1}{2!}C''\mathrm{d}x^2 + \cdots \tag{67}$$

where

$$C' = \frac{\partial C}{\partial x} \tag{68}$$

$$C'' = \frac{\partial^2 C}{\partial x^2} \tag{69}$$

The first term of Eq. (67) is

$$C'\mathrm{d}x = C'\left(v\mathrm{d}t + \sqrt{2D_B}\mathrm{d}W\right) \approx C'v\mathrm{d}t \tag{70}$$

using Eq. (60) of the time average, which implies that the diffusion time scale already satisfies the restricted condition of $\mathrm{d}t \gg \tau_p$. The second term of Eq. (67) is

$$C''(\mathrm{d}x)^2 = C''\left(v\mathrm{d}t + \sqrt{2D_B}\mathrm{d}W\right)^2 \approx C''2D_B\mathrm{d}t \tag{71}$$

after dropping the second order term of $\mathrm{d}t$ and the first order term of $\mathrm{d}W$. Substitution of Eqs. (70) and (71) with Eq. (67) gives

$$\mathrm{d}C(x) = C'v\mathrm{d}t + C''D_B\mathrm{d}t \tag{72}$$

and therefore

$$\frac{\partial C}{\partial t} = D_B\frac{\partial^2 C}{\partial x^2} + v\frac{\partial C}{\partial x} \tag{73}$$

which looks similar to the conventional convective diffusion equation with the sign of v reversed. Eq. (73) indicates that a group of identical particles of mass m undergoes convective and diffusive transport in the Eulerian space. A particle in the group is located at the position x at time t, moving with velocity v. This specific particle observes the concentration C of other particles nearby its position x. Therefore, Eq. (73) exists as the convective diffusion equation in the Lagrangian picture. If the particle moves with velocity v in a stationary fluid, then the motion is equivalent to particles that perform only diffusive motion within a fluid moving with $-v$. To emphasize the fluid velocity, we replace v with $-u$; then the Lagrangian convective diffusion Eq. (73) becomes the original (Eulerian) convection-diffusion equation:

$$\frac{\partial C}{\partial t} = D_B\frac{\partial^2 C}{\partial x^2} - u\frac{\partial C}{\partial x} \tag{74}$$

which can be directly obtained by replacing Eq. (65) by

$$\mathrm{d}x = -u\mathrm{d}t + \sqrt{2D_B}\mathrm{d}W \tag{75}$$

4. Dissipation rates

4.1 Energy consumption per time

In classical mechanics, work done due to an infinitesimal displacement of a particle $\mathrm{d}\boldsymbol{r}$ under the influence of force field $\boldsymbol{F}$ is

$$\mathrm{d}W = \boldsymbol{F}\cdot\mathrm{d}\boldsymbol{r} \tag{76}$$

The time differentiation of Eq. (76) provides an energy consumption rate (i.e., power represented by P) as a dot product of the particle velocity $\boldsymbol{v}$ and the applied force $\boldsymbol{F}$:

$$\dot{W} = \frac{\mathrm{d}W}{\mathrm{d}t} = \boldsymbol{v} \cdot \boldsymbol{F} \tag{77}$$

For an arbitrary physical quantity Q, variation rate of its density can be represented as

$$\frac{1}{V}\frac{\mathrm{d}Q}{\mathrm{d}t} = \frac{1}{V}\frac{\mathrm{d}\boldsymbol{r}}{\mathrm{d}t} \cdot \nabla Q = \boldsymbol{v} \cdot \nabla q \tag{78}$$

where V is the constant system volume and $q = Q/V$ is a volumetric density of Q or named specific Q. Eq. (78) indicates that a density changing rate of Q is equal to q operated by $\boldsymbol{v} \cdot \nabla$. If we replace Q by the internal energy of the system, then the specific energy consumption rate is expressed as

$$\dot{W} = \frac{1}{V}\frac{\mathrm{d}W}{\mathrm{d}t} = \boldsymbol{v} \cdot \nabla w = \frac{\boldsymbol{v}}{A_c} \cdot \nabla w' \tag{79}$$

where w and w' are specific work done and work done per length, respectively, and A_c is the cross-sectional area normal to $\nabla w'$. For a continuous media, $\nabla w'$ causes transport phenomena in a non-equilibrium state, and v/A_c is generated as proportional to a flux. A changing rate can be quantified as a product of a driving force and a flux, as implicated from Eq. (77).

Let us consider a closed system possessing ξ_1 and ξ_2, as some thermodynamic quantities characterizing the system state. The values of ξ_i at a state of equilibrium are denoted ξ_1^0 and ξ_2^0 and values outside equilibrium ξ_1' and ξ_2'. Within a static equilibrium, the entropy represented by S of the system is maintained as the maximum. For a system away from the static equilibrium, the generalized driving force is defined as

$$X_k = \nabla\left(\frac{\partial S}{\partial \xi_k}\right) \tag{80}$$

which is obviously zero for all k at the static equilibrium. A flux J_j of ξ_j is defined as

$$J_j = \frac{1}{A_c}\frac{\mathrm{d}\xi_j}{\mathrm{d}t} = \frac{\dot{\xi}_j}{A_c} = \sum_k L_{jk} X_k \tag{81}$$

which assumes that J_j represents a linear combination of all the existing driving forces X_k. We take Onsager's symmetry principle [12, 13], which indicates that the kinetic coefficient L_{jk} for all j and k are symmetrical such as

$$L_{jk} = L_{kj} \tag{82}$$

The entropy production rate per unit volume, or the specific entropy production rate, is defined as

$$\sigma = \frac{\mathrm{d}s}{\mathrm{d}t} \tag{83}$$

where $s = S/V$. We expand the specific entropy s with respect to infinitesimal changes of ξ_k as an independent variable:

$$\sigma = \sum_k \frac{\mathrm{d}\xi_k}{\mathrm{d}t}\frac{\partial s}{\partial \xi_k} = \sum_k (A_c J_k) \cdot \left(\frac{1}{A_c}\frac{\partial S}{\partial \xi_k}\right) = \sum_k J_k X_k \tag{84}$$

which represents the changing rate of the specific entropy as a dot project of flux $\boldsymbol{J}$ and driving force $\boldsymbol{X}$. The subscript k in Eq. (84) is for physical quantities on which the respective entropy depends. For mathematical simplicity, a new quantity is defined as $Y_k = TX_k$, where T is the absolute temperature in Kelvin to have

$$T\sigma = \sum_k J_k Y_k \tag{85}$$

Note that $T\sigma$ has a physical dimension equal to that of the specific power. An inverse relationship of Eq. (81) is

$$X_k = \sum_l R_{kl} J_l \tag{86}$$

where R_{kl} represents an inverse matrix of L_{jk}, i.e., $R_{kl}L_{jk} = \delta_{lj}$, which can be proven by substituting Eq. (86) with Eq. (81):

$$J_j = \sum_k L_{jk} X_k = \sum_l \left(\sum_k L_{jk} R_{kl} \right) J_l = \sum_l \delta_{jl} J_l = J_j \tag{87}$$

Substitution of Eq. (86) to Eq. (84) represents σ in terms of flux $\boldsymbol{J}$

$$\sigma = \sum_{j,\,k} J_j R_{jk} J_k = \langle \boldsymbol{J} | \boldsymbol{R} | \boldsymbol{J} \rangle \tag{88}$$

where $\langle \boldsymbol{J}|$ and $|\boldsymbol{J}\rangle$ exist as the row and column vectors of $\boldsymbol{J}$, respectively, and $\boldsymbol{R}$ represents the generalized resistance matrix. A partial derivative of σ with respect to an arbitrary flux J_i is equal to twice the generalized driving force:

$$\frac{\partial \sigma}{\partial J_i} = \sum_{j,\,k} \left[\delta_{ij} R_{jk} J_k + J_j R_{jk} \delta_{ik} \right] = \sum_k [R_{ik} J_k] + \sum_j \left[J_j R_{ji} \right] = 2X_i \tag{89}$$

In this case, the specific entropy dissipation rate σ is presented using the flux and σ differential with the flux by substituting Eq. (89) with Eq. (84):

$$\sigma = \frac{1}{2} \sum_k J_k \frac{\partial \sigma}{\partial J_k} \tag{90}$$

which indicates that the specific entropy increases with respect to the flux and proves that the systems is away from a pure, static equilibrium state.

4.2 Effective driving forces

The second thermodynamic law represents the infinitesimal entropy change in the microcanonical ensemble:

$$\mathrm{d}S = \frac{1}{T}\mathrm{d}E + \frac{P}{T}\mathrm{d}V - \sum_i \left(\frac{\mu_i}{T} \right) \mathrm{d}N_i \tag{91}$$

where E represents the internal energy, P represents the system pressure within a volume V, and μ_i and N_i are the chemical potential and the mole number of species i, respectively. Eq. (91) implies the entropy $S = S(E, V, N_i)$ as a function of the internal energy E, the volume V, and the number of species i N_i.This gives $\xi_1 = E$, $\xi_2 = V$, and $\xi_i = N_i$ ($i = 3$ for water and $i = 4$ for solute). The driving forces X_i are particularly calculated as

$$X_1 = X_q = \nabla\left(\frac{\partial S}{\partial E}\right)_{V, N_s} = \nabla\left(\frac{1}{T}\right) \tag{92}$$

$$X_2 = X_v = \nabla\left(\frac{\partial S}{\partial V}\right)_{E, N_s} = \nabla\left(\frac{P}{T}\right) \tag{93}$$

$$X_3 = X_s = \nabla\left(\frac{\partial S}{\partial N_s}\right)_{E, V} = \nabla\left(-\frac{\mu_s}{T}\right) \tag{94}$$

where subscripts q, v, and s of X indicates heat, volume of solvent, and solute, respectively. In Eq. (92), entropy S is differentiated by energy E, keeping V, and N_s invariant, which are applied to Eqs. (93) and (94). Eq. (94) indicates that the driving force is a negative gradient of the chemical potential divided by the ambient temperature. Within the isothermal-isobaric ensemble, Gibbs free energy is defined as

$$G = H - TS \tag{95}$$

where $H(= E + PV)$ is enthalpy. If the solute concentration is diluted (i.e., $N_w \gg N_s$), it is referred to as a weak solution. As such, the overall chemical potential can be approximated as

$$\mu = \frac{\partial G}{\partial N} = \frac{G}{N_w + N_s} \simeq \frac{G}{N_w} = \mu_w = \overline{H} - \overline{S}T \tag{96}$$

where $\overline{H}$ and $\overline{S}$ represent molar enthalpy and entropy, respectively. An infinitesimal change of Gibbs free energy is, in particular, written as

$$\mathrm{d}G = -S\mathrm{d}T + V\mathrm{d}P + \mu_s \mathrm{d}N_s \tag{97}$$

which is equivalent to

$$\mathrm{d}\left(\frac{G}{N_w}\right) \simeq \mathrm{d}\mu_w = -\overline{S}\mathrm{d}T + \overline{V}\mathrm{d}P + \mu_s \mathrm{d}c \tag{98}$$

where $\overline{V}$ is a molar volume of the system, μ_s is the solute chemical potential, and $c = N_s/N_w$ is the molar fraction of solute molecules. The gradient of the solvent chemical potential was rewritten as a linear combination of gradients of temperature, pressure, and molar solute fraction:

$$\nabla\mu_w = -\overline{S}\nabla T + \overline{V}\nabla P + \mu_s \nabla c \tag{99}$$

where the following mathematical identity was used

$$\nabla\frac{\mu_k}{T} = \mu_k \nabla\left(\frac{1}{T}\right) + \frac{1}{T}(\nabla\mu_k) \tag{100}$$

In general, fluxes of heat, solvent volume, and solute molecules are intrinsically coupled to their driving forces, such as

$$\begin{bmatrix} J_q \\ J_v \\ J_s \end{bmatrix} = \begin{bmatrix} L_{qq} & L_{qv} & L_{qs} \\ L_{qv} & L_{vv} & L_{vs} \\ L_{qs} & L_{vs} & L_{ss} \end{bmatrix} \begin{bmatrix} X_q \\ X_v \\ X_s \end{bmatrix} \tag{101}$$

where Onsager's reciprocal relationship, $L_{ij} = L_{ji}$, is employed.

4.3 Applications

4.3.1 Solute diffusion

The primary driving force for the solute transport is $X_s = -\nabla(\mu_s/T)$, if temperature and pressure gradients are not significant in solute transport. We consider the diffusive flux of solute only in an isothermal and isobaric process and neglect terms of L_{qs} and L_{vs}:

$$J_s = -L_{ss}\nabla\frac{\mu_s}{T} = -\frac{L_{ss}}{T}\nabla\mu_s \tag{102}$$

which is equivalent to Fick's law of

$$J_s = -D\nabla c \tag{103}$$

where D [m^2/s] is a solute diffusion coefficient. If Eq. (102) is expressed in terms of concentration gradient, we have

$$J_{ss} = -\frac{L_{ss}}{T}\left(\frac{\partial\mu_s}{\partial c}\right)\nabla c \tag{104}$$

By Eqs. (103) and (104), one can find

$$\frac{L_{ss}}{T}\left(\frac{\partial\mu_s}{\partial c}\right)_T = D \tag{105}$$

Then, the entropy-changing rate based on the solute transport is calculated as

$$\sigma_s = J_s X_s = \frac{L_{ss}}{T^2}(\nabla\mu_s)^2 = \frac{D/T}{(\partial\mu_s/\partial c)_T}(\nabla\mu_s)^2 \tag{106}$$

Next, we consider the Stokes-Einstein diffusivity:

$$D_{SE} = \frac{k_B T}{3\pi\eta d_p} \tag{107}$$

where k_B is Boltzmann constant, η is the solvent viscosity, and d_p is the diameter of a particle diffusing within the solvent medium. The phenomenological coefficient L_{qq} is represented as

$$L_{ss} = \frac{D_{SE}T}{(\partial\mu_s/\partial c)_T} = \frac{k_B T}{3\pi\eta d_p}\cdot T\left(\frac{\partial\mu_s}{\partial c}\right)_T^{-1} \tag{108}$$

For weakly interacting solutes, the solute chemical potential is

$$\mu = \mu_0 + \mathcal{R}T\ln a \tag{109}$$

where μ_0 is generally a function of T and P, which are constant in this equation, $\mathcal{R}$ is the gas constant, and a is the solute activity. For a dilute solution, the activity represented by a is often approximated as the concentration c (i.e., $a \simeq c$). The proportionality between L_{ss} and D_{SE} is

$$T\left(\frac{\partial \mu_s}{\partial c}\right)_T^{-1} = T\frac{1}{\mathcal{R}T/c} = \frac{c}{\mathcal{R}} \tag{110}$$

which leads to

$$L_{ss} = \frac{k_B T}{3\pi\eta d_p}\frac{c}{\mathcal{R}} = \frac{N_A Tc}{3\pi\eta d_p} \tag{111}$$

where N_A is the Avogadro constant.

For a dilute isothermal solution, we represent the entropy-changing rate as

$$\sigma_s = D\mathcal{R}\frac{(\nabla c)^2}{c} = \frac{\mathcal{R}}{D}\frac{J_{ss}^2}{c} \tag{112}$$

for an isothermal and isobaric process. Assuming that D is not a strong function of c, Eq. (112) indicates that the diffusive entropy rate σ_s is unconditionally positive (as expected), increases with the diffusive flux, and decreases with the concentration c. Within this analysis, c is defined as molar or number fraction of solute molecules to the solvent. For a dilute solution, conversion of c to a solute mass or mole number per unit volume is straightforward.

4.3.2 Thermal flux

The thermal flux consists of conductive and convective transports, proportional to ∇T and ∇P, respectively. Neglecting the solute diffusion in Eq. (101), the coupled equations of heat and fluid flows are simplified as

$$\begin{bmatrix} J_q \\ J_v \end{bmatrix} = \begin{bmatrix} \alpha & \beta \\ \beta & \gamma \end{bmatrix}\begin{bmatrix} X_q \\ X_v \end{bmatrix} \tag{113}$$

using Onsager's reciprocal relationship in that the off-diagonal coefficients are symmetrical. The driving forces are

$$\begin{bmatrix} X_q \\ X_v \end{bmatrix} = \begin{bmatrix} \nabla 1/T \\ \nabla P/T \end{bmatrix} = \frac{-1}{T^2}\begin{bmatrix} 1 & 0 \\ P & -T \end{bmatrix}\begin{bmatrix} \nabla T \\ \nabla P \end{bmatrix} \tag{114}$$

The substitution of Eq. (113) into (114) gives

$$\begin{aligned} -T^2\begin{bmatrix} J_q \\ J_v \end{bmatrix} &= \begin{bmatrix} \alpha & \beta \\ \beta & \gamma \end{bmatrix}\begin{bmatrix} 1 & 0 \\ P & -T \end{bmatrix}\begin{bmatrix} \nabla T \\ \nabla P \end{bmatrix} \\ &= \begin{bmatrix} \alpha + P\beta & -T\beta \\ \beta + P\gamma & -T\gamma \end{bmatrix}\begin{bmatrix} \nabla T \\ \nabla P \end{bmatrix} \end{aligned} \tag{115}$$

or

$$-T^2\begin{bmatrix} J_q/\beta \\ J_v/\gamma \end{bmatrix} = \begin{bmatrix} \alpha/\beta + P & -T \\ \beta/\gamma + P & -T \end{bmatrix}\begin{bmatrix} \nabla T \\ \nabla P \end{bmatrix} \tag{116}$$

Subtracting the first row by the second row of Eq. (116) provides

$$\left(\frac{J_q}{\beta} - \frac{J_v}{\gamma}\right) = -\frac{1}{T^2}\left(\frac{\alpha}{\beta} - \frac{\beta}{\gamma}\right)\nabla T \tag{117}$$

$$J_q = \frac{\beta}{\gamma} J_v - \frac{1}{T^2}\left(\alpha - \beta\frac{\beta}{\gamma}\right)\nabla T \tag{118}$$

Through physical interpretation, one can conclude that

$$\frac{\beta}{\gamma} = \tilde{h} \tag{119}$$

where $\tilde{h}$ represents the system enthalpy as a function of temperature. Finally, the coupled heat transfer equation is

$$J_q = -\kappa_q \nabla T + \tilde{h} J_v \tag{120}$$

where

$$\kappa_q = \alpha - \beta\frac{\tilde{h}}{T^2} \tag{121}$$

is the thermal conductivity.

5. Concluding remarks

In this chapter, we investigated diffusion phenomenon as an irreversible process. By thermodynamic laws, entropy always increases as a system of interest evolves in a non-equilibrium state. The entropy-increasing rate per unit volume is a measure of how fast the system changes from the current to a more disordered state. Entropy concept is explained from the basic mathematics using several examples. Diffusion phenomenon is explained using (phenomenological) Fick's law, and more fundamental theories were summarized, which theoretically derive the diffusion coefficient and the convection-diffusion equation. Finally, the dissipation rate, i.e., entropy-changing rate per volume, is revisited and obtained in detail. The coupled, irreversible transport equation in steady state is applied to solute diffusion in an isothermal-isobaric process and heat transfer that is consisting of the conductive and convective transport due to the temperature gradient and fluid flow, respectively. As engineering processes are mostly open in the steady state, the theoretical approaches discussed in this chapter may be a starting point of the future development in irreversible thermodynamics and statistical mechanics.

Author details

Albert S. Kim
University of Hawaii at Manoa, USA

*Address all correspondence to: albertsk@hawaii.edu

References

[1] Kim AS, Kim H-J. Membrane thermodynamics for osmotic phenomena. In: Yonar T, editor. Desalination. Rijeka: InTechOpen; 2017. pp. 1-26. ISBN: 978-953-51-3364-3

[2] Shannon CE. A mathematical theory of communication. Bell System Technical Journal. 1948;**270**(3):379-423. DOI: 10.1002/j.1538-7305.1948.tb01338.x

[3] Fick A. Ueber diffusion. Annalen der Physik und Chemie. 1855;**170**(1):59-86

[4] Fick A. On liquid diffusion. Journal of Membrane Science. 1995;**100**(1): 33-38. ISSN: 0376-7388. DOI: 10.1016/ 0376-7388(94)00230-v

[5] Stokes GG. On the effect of internal friction of fluids on the motion of pendulums. Transactions of the Cambridge Philosophical Society. 1851;**9**:1-106

[6] Einstein A. Investigations of the Theory of the Brownian Movement. New York, NY: Dover Publications, Inc.; 1956

[7] Langevin P. Sur la theorie du mouvement brownien. Comptes Rendus de l'Académie des Sciences (Paris). 1908;**146**:530-533

[8] Howard Lee M. Solutions of the generalized Langevin equation by a method of recurrence relations. Physical Review B. 1982;**260**(5):2547-2551

[9] Wiener N. Differential Space. Journal of Mathematical Physics. 1923;**58**:31-174

[10] Ito M. An extension of nonlinear evolution equations of the K-dV (mK-dV) type to higher orders. Journal of the Physical Society of Japan. 1980;**49**(0): 771-778. DOI: 10.1143/JPSJ.49.771

[11] Gardiner CW. Handbook of Stochastic Methods: For Physics, Chemistry and the Natural Sciences. Springer; 1996

[12] Onsager L. Reciprocal relations in irreversible processes. II. Physical Review. 1931;**38**(12):2265-2279. DOI: 10.1103/physrev.38.2265

[13] Onsager L. Reciprocal relations in irreversible processes. I. Physical Review. 1931;**37**(4):405-426. DOI: 10.1103/PhysRev.37.405

Section 2

Non-Equilibrium Simulations

Chapter 4

Using the Principles of Nonequilibrium Thermodynamics for the Analysis of Phase Transformations in Iron-Carbon Alloys

Bobyr Sergiy Volodimyrovych

Abstract

Using the principles of nonequilibrium thermodynamics, a technique has been developed for calculating diffusion flows during phase transformations in iron-carbon alloys. Expressions for the calculation of cross coefficients, driving forces, and flows in Onsager equations for the model thermodynamic system are given; examples of the use of the developed technique are given for the processes of graphitization and the formation of carbides in chromium steel during tempering. The nonequilibrium thermodynamics analysis of the eutectoid transformation is executed into carbon steel. Onsager's equations of motion are built for the model thermodynamics system describing eutectoid transformation. The basic kinetic parameters of process are growth rate of perlite and between inter-plates distance for the stationary process of eutectoid transformation. We founded dependencies of basic kinetic parameters of process from the size of supercooling. A nonequilibrium thermodynamic model of the austenite nondiffusion transformation in iron and alloys based on it is developed, taking into account internal stresses in the system. Onsager motion equations are found for a model thermodynamic system describing a nondiffusion transformation and kinetic equations for changing deformations and growth rates of the α-phase. A scheme of austenitic nondiffusion transformations is constructed, including normal and martensitic transformations, as limiting cases.

Keywords: nonequilibrium thermodynamics, the iron-based alloys, transformation of austenite, diffusion, equations of motion, nondiffusion transformation

1. Introduction

The study of phase transformations is one of the most important problems in the physics of metals [1–3]. Phase transformations are divided into diffusion and nondiffusion [1]. If the kinetics of phase transformation in steels and cast irons is determined by the diffusion of carbon, this allows them to be attributed to conversions controlled by diffusion [1–4]. Such transformations in iron-carbon alloys include pearlitic transformation of austenite, and transformations occurring during tempering, graphitization of undoped cementite, separation of carbides in alloyed steels, and others [4–6].

When the rate of transformation of austenite is determined by the rate at which the interface separates, differing only in its crystalline structure, the transformation is called nondiffusion [1]. Kinetically, the normal polymorphic and martensitic transformations of austenite are distinguished. When the temperature of the normal transformation decreases, its velocity first increases and then decreases. The kinetics of the martensitic transformation is characterized by a very high rate of growth of individual crystals and the maximum space velocity at the initial moment of transformation under isothermal conditions.

In addition to martensite, at least two other structural components are known, which arc formed with a shear ("martensitic") morphology of crystal formation—ferrite side-plates and acicular ferrite. They can also be attributed, with some simplifying assumptions, to the products of the nondiffusion transformation of austenite. In addition, in some alloys martensitic and normal transformations occur at the same temperature [1]. The consistent theory of nondiffusion transformations should explain this phenomenon. Thus, the theoretical description of the processes of phase transformations in iron-carbon alloys is a complex and urgent task of modern metal physics.

Nonequilibrium thermodynamics provides the necessary apparatus for analyzing the processes of phase transformations in iron-carbon alloys [7–9]. In the general case, the thermodynamic equations of motion have the form [7]:

$$J_i = \sum_{\kappa=1}^{N} L_{ik} X_k (\mathrm{i} = 1, .., \mathrm{N}), \tag{1}$$

where J_i are flows, X_k are the thermodynamic forces, $L_{ik} = L_{ki}$ are the Onsager kinetic coefficients [9], and i, k are the charge numbers (transfer substrates).

The main driving forces of phase transformations in nonequilibrium thermodynamics are gradients of the chemical potentials of their components [6–9]. When discontinuous systems are considered, the finite differences of chemical potentials $(-\Delta\mu_i,)$ as the transition from a metastable state to a stable state are used as thermodynamic forces [10, 11]. Equations of nonequilibrium thermodynamics were first used in the physics of metals to describe the process of graphitization of nonalloyed iron-carbon alloys [6, 11].

As is known, unalloyed cementite in iron-carbon alloys at normal pressure is a metastable phase, its activity in phases with it in equilibrium exceeds the solubility of graphite, a stable phase [11]. Therefore, at a sufficiently high temperature, graphitization of such alloys takes place, that is, phase transition from metastable to stable equilibrium. Despite the seeming simplicity of this process, its theoretical description is a complex task.

If two values are used as charges of the graphitization process-carbon and iron concentrations, then, according to (1), the equations of motion take the form:

$$J_1 = L_{11} X_1 + L_{12} X_2 \tag{2}$$

$$J_2 = L_{21} X_1 + L_{22} X_2, \tag{3}$$

where J_1 is the carbon flow characterizing the rate of the graphitization process, J_2 is the flow of iron, and $X_1 = (-\Delta\mu_{Fe})$ and $X_2 = (-\Delta\mu_C)$ are the thermodynamic forces of iron and carbon. The potential drop has a "+" sign as it increases, and the flow is directed toward a decrease in the potential, so the expressions for the forces contain the sign "–."

The main question that must be solved when using the Onsager Eqs. (1)–(3) is the values of the cross coefficients.

In [5], for the first time on the basis of a special variational procedure, an expression for the cross coefficients in the Onsager equations was proposed in the form:

$$L_{21} = L_{12} = -\sqrt{L_{11} \times L_{22}}, \tag{4}$$

and the sign–before the root is chosen on the basis that the observed flux of iron with respect to the flow of carbon had a negative sign.

As shown in [6, 11], in the complex process with two flows, an increase in the potential of one of the charges is observed, that is, one process is "leading," and the other is "driven." The "driven" process in itself, i.e., in isolation from the "lead," is not possible, since thermodynamically not beneficial. In the system of Eqs. (2) and (3), the thermodynamic force ($-\Delta\mu_{Fe}$) is negative and inhibits the process as a whole, the diffusion of iron is a forced process, and the leading one is the diffusion of carbon.

Thus, the graphitization process must be accompanied by a very intensive transfer of a solid solution (mainly iron), which makes it possible for the phase with a low-density graphite to grow in it. The authors of [6, 11] assumed that the factor contributing to graphitization is the pressure that arises in the austenite matrix under the action of graphite inclusions that expand it. However, in [12], considering the mechanism of graphitization of cast irons during thermocyclic treatment, K.P. Bunin with AA. Baranov came to the conclusion that the absolute value of the contact pressures is an order of magnitude less than the necessary for the dislocation creep mechanism under the influence of contact pressure. Since graphite films in pores cannot possess super strong properties, the evacuation of matrix atoms is apparently carried out by another mechanism.

In [5], using nonequilibrium thermodynamic methods, it was shown that under the conditions of the system's striving for dynamic equilibrium, the concentration of vacancies in graphite inclusion becomes less than the vacancy concentration at the γ-phase-graphite boundary. This can occur as a result of approaching the γ-phase boundary—graphite of austenitic vacancies. In this case, the thermodynamic force ($-\Delta_{\mu 0 v}$) prevents the graphitization, and the reduced graphitization force ($-\Delta_{\mu^* C}$) decreases to zero and can even take a negative value.

Thus, the goal of this paper is to show how the methods of nonequilibrium thermodynamics can be used fruitfully to solve the theoretical problems of metal physics, namely, the analysis of phase transformations. Let us further consider the application of the principles of nonequilibrium thermodynamics to the analysis of specific cases of phase transformations in iron-carbon alloys.

2. Formation of carbides in chrome steel during tempering

Consider the process of separation of carbides in a low-carbon steel system of iron-carbon-chromium with 0.15% carbon and about 5% chromium at 600°C. In this model system, there are two phases—the doped α-phase (F) and carbides (K), in which carbon, iron, chromium, and vacancies flows (**Figure 1**). As charges, we will use four quantities—the concentrations of carbon, iron, chromium, and vacancies. The flow of vacancies in the carbide phase will be assumed to be equal to the flow of vacancies in the ferrite.

In the absence of a change in the volume of the system, for flows in the doped α phase, condition [13] is fulfilled:

$$J_{Fe} + J_{Cr} + J_v = 0, \tag{5}$$

so one of the threads (in our case—J_v) is a dependent quantity. According to (1), the thermodynamic equations for flows in the carbide phase take the form:

$$J_{Fe} = -L_{11}\Delta\mu_{Fe} - L_{12}\Delta\mu_c - L_{13}\Delta\mu_{Cr} \tag{6}$$

$$J_C = -L_{21}\Delta\mu_{Fe} - L_{22}\Delta\mu_c - L_{23}\Delta\mu_{Cr} \tag{7}$$

$$J_{Cr} = -L_{31}\Delta\mu_{Fe} - L_{32}\Delta\mu_c - L_{33}\Delta\mu_{Cr}, \tag{8}$$

where J_{Fe}, J_C, and J_{Cr} are the flows of iron, carbon, and chromium, respectively.

Based on the general principles of nonequilibrium thermodynamics, we can find the values of the thermodynamic forces $-\Delta\mu_{Fe}$, $\Delta\mu_{Cr}$, and $-\Delta\mu_C$, as well as the values of the kinetic coefficients L_{12}, L_{13}, and L_{23}, as it was previously performed in [5] for a system with two flows. In the conditions of complete equilibrium, $\Delta\mu_{Fe} = 0$, $\Delta\mu_C = 0$, and $\Delta\mu_{Cr} = 0$. However, for a linear thermodynamic system, there is also the possibility of dynamic equilibrium, in which all the flows are 0, but some thermodynamic forces in the system are not equal to zero (there are their variations) [5, 7].

Let us consider this possibility for a triple thermodynamic system. From Eqs. (6)–(8), it follows that near equilibrium, in the presence of variations of thermodynamic forces, the following conditions must be fulfilled:

$$J_{Fe} = 0 \Rightarrow L_{11}\delta\mu_{Fe} + L_{12}\delta\mu_C + L_{13}\delta\mu_{Cr} = 0, \tag{9}$$

$$J_C = 0 \Rightarrow L_{21}\delta\mu_{Fe} + L_{22}\delta\mu_C + L_{13}\delta\mu_{Cr} = 0, \tag{10}$$

$$J_{Cr} = 0 \Rightarrow L_{31}\delta\mu_{Fe} + L_{32}\delta\mu_C + L_{33}\delta\mu_{Cr} = 0, \tag{11}$$

where the index $\delta\mu$ denotes the coordinated variations of the thermodynamic forces that ensure the dynamic equilibrium of the system. It follows from the system of Eqs. (9)–(11) that the expressions for the flows of iron, chromium, and carbon are connected: the cross rates L_{12}, L_{13}, and L_{23} in expressions for the flows must have values such that the determinant of the matrix A composed of the coefficients of this system was equal to 0. In this case, the values of the flows of iron and chromium can significantly increase due to cross-kinetic coefficients in comparison with the independent diffusion of these elements [7, 16].

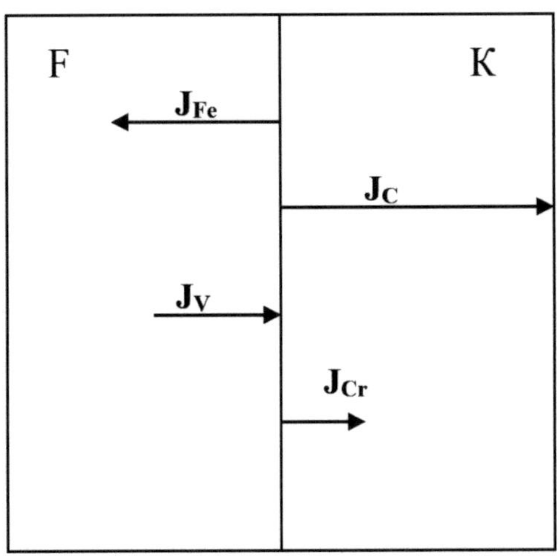

Figure 1.
Scheme of the process of carbides formation in chromium steel.

Let us find the expressions for the cross coefficients, which make it possible to obtain a nontrivial solution of the system of Eqs. (9)–(11). From the first Eq. (9), we establish a connection between the variations of forces:

$$\delta\mu_{Fe} = -(L_{12}/L_{11})\delta\mu_C - (L_{13}/L_{11})\delta\mu_{Cr}. \tag{12}$$

Substituting (12) into Eqs. (10) and (11), we find

$$J_C = \left(L_{22} - L_{12}^2/L_{11}\right)\delta\mu_C + (L_{23} - L_{12}L_{13}/L_{11})\delta\mu_{Cr} = 0, \tag{13}$$

$$J_{Cr} = (L_{32} - L_{13}L_{12}/L_{11})\delta\mu_C + \left(L_{33} - L_{13}^2/L_{11}\right)\delta\mu_{Cr} = 0, \tag{14}$$

For independent variations $\delta\mu_C$ and $\delta\mu_{Cr}$, the linear system of Eqs. (13) and (14) is compatible if the coefficients of $\delta\mu_C$ and $\delta\mu_{Cr}$ are equal to 0, from which we immediately find the relation between Onsager's kinetic coefficients:

$$L_{ik} = L_{ki} = \pm\sqrt{L_{ii} \times L_{kk}},\ \text{i, k} = 1...3 \tag{15}$$

and the sign before the root is selected based on the sign (direction) of the flows under consideration (see **Figure 1**). The considered procedure of variation allows us to find cross rates in the Onsager equations after the direct kinetic coefficients are calculated. In this case, the established connection (15) is satisfied for systems not very far from equilibrium and for the real system is approximate.

3. Calculation of thermodynamic forces and kinetic coefficients

Let us find the values of the thermodynamic forces and kinetic coefficients for the steel of the Fe-C-Cr system with 0.15% C at 600°C. We will assume that in a solid α-solution, there is chromium with a concentration of $C_{Cr} = 0.05$ and a carbon with a concentration of $C_C = 0.007$, an iron concentration of $C_{Fe} = 0.943$. In cementite-type carbide, chromium is found with a mass fraction of ~20% (with a concentration of $C_{Cr} = 0.2$) and carbon with a carbon concentration of 0.25, an iron concentration in the carbide $C'_{Fe} = 0.55$.

It is known from the experimental data that carbon is removed very rapidly (approximately 1 minute) from the α-solution of alloyed steel at a temperature of 550–650°C and, consequently, the formation of carbide inclusions is primarily due to carbon diffusion [14].

The thermodynamic force for carbon can be calculated from the formula [11]:

$$-\Delta\mu_C = -RT\ln\frac{a_C^K}{a_C^\alpha}, \tag{16}$$

where $a^\alpha{}_C$ is the thermodynamic activity of carbon in α-solution, $a^K{}_C$ is the thermodynamic activity of carbon in cementite, R is the universal gas constant, and T is the temperature of the alloy.

The change in the thermodynamic activity of carbon in the alloy upon doping with component i can be found by the method of [15, 16] from the equation:

$$\ln\left(a_C/a_{CO}\right) = \beta i\ Ni, \tag{17}$$

where βi is the coefficient of the element's influence on the thermodynamic activity of carbon in the alloy, Ni is the content of the element in the alloy in atomic fractions, and a_{CO} is the thermodynamic activity of carbon for the alloy in the standard state.

We will assume that for our steel in the standard state $a^{\alpha}{}_{CO} = a^{K}{}_{CO} = a_{CO}$, i.e., unalloyed cementite in steel with 0.15%, C is stable and in equilibrium with the solid solution at a tempering temperature of 600°C [13]. Using this condition and Eqs. (16) and (17), we find:

$$\ln\left(a^{K}{}_{C}/a^{\alpha}{}_{C}\right) = \beta^{K}{}_{Cr}\, N^{K}{}_{Cr} - \beta^{\alpha}{}_{Cr}\, N^{\alpha}{}_{Cr} \tag{18}$$

The value of βi is calculated through the interfacial distribution coefficient of the alloying element $Ki = Ni\,(K)/Ni\,(\alpha)$ and the atomic fraction of carbon in the alloy Nc [15, 16]:

$$\beta i = -\frac{(Ki - 1) + (Nc(K) - KiNc(\alpha))}{(Ki - 1)Nc + (Nc(K) - KiNc(\alpha))}. \tag{19}$$

With a slight error for low-alloyed alloys, we can take $Nc\,(K) = 0.25$, $Nc\,(\alpha) \approx 0.001$—the carbon content in the undoped phases of steel, taken from the Fe-C state diagram.

Using the coefficient of chromium distribution between the α-phase and the carbide K_{Cr}, equal to 4, we find the equations for calculating the coefficients of influence β_{Cr}:

$$\beta_{Cr} = -3.246/(3{,}0\mathrm{Nc} + 0.246),$$

whence $\beta^{\alpha}{}_{Cr} = -12.16$ and $\beta^{K}{}_{Cr} = -3.26$.

Then from expressions (16)–(18), one can find the values.

$$\ln\left(a^{K}{}_{C}/a^{\alpha}{}_{C}\right) = -0.6085 + 0.652 = -0.0425 \text{ and } - \Delta_{\mu C} = 308.47 \text{ Joule}. \tag{21}$$

The work done in the diffusion of carbon from the α-phase to cementite is positive. For the diffusion of iron, it is not possible to calculate the difference of thermodynamic potentials, since the coefficient of iron activity in carbide is unknown. However, from the experimental data and the thermodynamics of the process, it is known that diffusion of carbon is the leading one, the diffusion of chromium accompanies the diffusion of carbon, and the diffusion of iron is forced, since it is directed toward increasing the concentration of iron.

With this in mind, we find the values of the kinetic coefficients L_{ii} in the Onsager equations.

As is known [8, 13], the kinetic coefficients L_{ii} are related to the diffusion coefficients D_i by the relation:

$$L_{ii} = C_i D_i / RT, \tag{22}$$

where C_1 is the concentration of iron in the alloy (0.943), C_2 is the concentration of carbon in the alloy (0.007), and C_3 is the concentration of chromium in the alloy (0.05).

Dependences of the diffusion coefficients of chromium and carbon in doped chromium ferrite on the temperature have the form [14, 17]:

$$D^{\alpha}_{C} = 0{,}177 \exp\left[\frac{-88230}{RT}\right] см^2/сек, \tag{23}$$

$$D^{\alpha}_{Fe} = 2{,}910^{-4} \exp\left[\frac{-251000}{RT}\right] см^2/сек, \tag{24}$$

$$D^{\alpha}_{Cr} = 3,05\exp\left[\frac{-358000}{RT}\right] см^2/сек. \tag{25}$$

At a temperature of 600° C:

$D_1 = D^{\alpha}_{Fe} \approx 3.03 \cdot 10^{-19}$ cm^2/s, $D_2 = D^{\alpha}_{C} \approx 1.02\ 10^{-6}$ cm^2/s, and $D_3 = D^{\alpha}_{Cr} \approx 1.38\ 10^{-21}$ cm^2/s.

Using relations (23)–(25) and (15), we find the values of the kinetic coefficients for our system:

$L_{11} = 0.394 \times 10^{-22}$, $L_{22} = 0.984 \times 10^{-13}$, $L_{12} = -1.97 \times 10^{-17}$, $L_{33} = 0.95 \times 10^{-26}$, $L_{13} = -0.611 \times 10^{-24}$, and

$L_{23} = 0.306 \times 10^{-19}$. Consequently, the system of Eqs. (6)–(8) takes the form:

$$J_{Fe} = 0.394 \times 10^{-22}(-\Delta\mu_{Fe}) - 1.97 \times 10^{-17}(-\Delta\mu_C) - 0.611 \times 10^{-24}(-\Delta\mu_{cr}), \tag{26}$$

$$J_C = -1.97 \times 10^{-17}(-\Delta\mu_{Fe}) + 0.984 \times 10^{-13}(-\Delta\mu_C) + 0.306 \times 10^{-19}(-\Delta\mu_{Cr}), \tag{27}$$

$$J_{Cr} = -0.611 \times 10^{-24}(-\Delta\mu_{Fe}) + 0.306 \times 10^{-19}(-\Delta\mu_C) + 0.95 \times 10^{-26}(-\Delta\mu_{Cr}). \tag{28}$$

It follows from Eqs. (26)–(28) that the values of iron and chromium fluxes increase substantially due to the cross-ratios L_{12} and L_{32} of a significant thermodynamic force ($-\Delta\mu_C$). The value of the carbon flux having a positive sign is determined mainly by the intrinsic coefficient L_{22}. The thermodynamic forces of iron and chromium make an insignificant contribution to the fluxes, because of the small value of the kinetic coefficients and their influence can be neglected. Then, as direct calculations show:

$J_{Fe} = -\ 6.08 \times 10^{-15}$, $J_C = 3.04 \times 10^{-11}$, $J_{Cr} = 0.94 \times 10^{-17}$, and

$J_V = 6.07 \times 10^{-15}$ cm^2/s.

It was established in [18] that during the tempering period, a certain amount of nanoparticles of special chromium carbide with a size of ~100 nm can be formed in the steel, which were detected experimentally.

4. The nonequilibrium thermodynamics analysis of the eutectoid transformation

In [19], a generalization of the equations characterizing the growth of the pearlite colony is proposed, based on the application of nonequilibrium thermodynamic methods.

To this end, Eq. (19) from [20], which characterizes the growth rate of a perlite colony, is represented as:

$$\frac{dX}{dt} = D_x\left[\left(C'_{ф}-C'_{ц}\right)/\left(C'_{ф}-C_{ф}\right) + \left(C'_{ф}-C'_{ц}\right)/\left(C_{ц}-C'_{ц}\right)\right]/\Delta = (D_x/\Delta)(-\Delta\varphi), \tag{29}$$

where D_x is the carbon diffusion coefficient in austenite along the x axis at a given temperature T, Δ is the thickness of a layer of austenite with different concentration of carbon, C'_{Φ} and C'_y is the carbon concentration in the austenite near the ferrite and cementite plates, respectively, at a temperature T (**Figure 2**), C_y is the carbon content in cementite (~6.67%), C_{Φ} is the carbon content in the ferrite at a given temperature T, and $-\Delta\varphi$ is the thermodynamic force of perlite lateral growth. It is determined by the carbon concentrations in ferrite and cementite and has a dimensionless value.

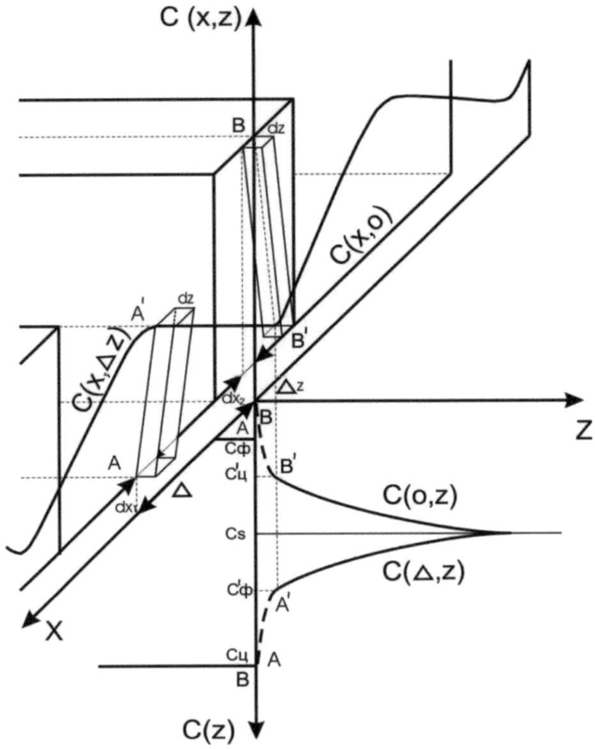

Figure 2.
Carbon distribution in the austenite-perlite system [20].

The second equation characterizing our system—the heat balance Eqs. (23)–(25) from [20]–is written in the form:

$$C\gamma dT/dt = \alpha\,\Delta T - (q\gamma/\Delta)\,dX/dt \tag{30}$$

where α is the heat transfer coefficient, C is the specific heat, ΔT is the temperature difference between the sample (T) and the cooling medium, q is the specific amount of heat released during the formation of perlite, and γ is the density of steel.

If two quantities are used as charges for the eutectoid transformation of austenite-the temperature of the sample T and the thickness of the plates of perlite X, then, according to (4), the Onsager motion equations must have symmetric forms (2) and (3),

where $J_1 = -\,dX/dt$ is the flow of the pearlitic layer (with increasing absolute value of the thermodynamic growth force of perlite, the flow increases in absolute value), and $J_2 = -\,C\gamma dT/dt$ is the heat flow in the sample (with a drop in sample temperature, the flow is positive), $X_1 = (-\Delta\varphi)$, $X_2 = (-\Delta T/T)$ is the thermodynamic forces of perlite growth and temperature [14].

In order for Eq. (29) to correspond to Eq. (2), it must contain an additional term $L_{12}\,(-\Delta T/T)$; with the value of the coefficient $L_{11} = (D_x/\Delta)$:

$$J_1 = (D_x/\Delta)(-\Delta\varphi) + L_{12}(-\Delta T/T) \tag{31}$$

where L_{12} is a cross ratio whose value is not yet known. Thus, we introduce (we assume) an additional dependence of the growth rate of the perlite layer not only on the carbon concentrations in the phases but also on the temperature.

Substituting expression (34) into the energy balance Eq. (33), we find the expression for the heat flow J_2:

$$J_2 = (q\gamma/\Delta)((D_x/\Delta)(-\Delta\varphi) + L_{12}(-\Delta T/T)) - \alpha T(-\Delta T/\mathrm{T})$$
$$= (q\gamma D_x/\Delta^2)(-\Delta\varphi) + (q\gamma L_{12}/\Delta - \alpha T)\ (-\Delta T/\mathrm{T}) \quad (32)$$

Relating Eqs. (3) and (32) to each other, we obtain:

$$L_{21} = q\gamma D_x/\Delta^2 \quad (33)$$

Using for the kinetic coefficients, the Onsager reciprocity relations $L_{ik} = L_{ki}$ [9], we find that

$$L_{12} = L_{21} = q\gamma D_x/\Delta^2, \quad (34)$$

whereas

$$L_{22} = q^2\gamma^2 D_x/\Delta^3 - \alpha T. \quad (35)$$

The system of Eqs. (31) and (32) takes the form:

$$J_1 = (D_x/\Delta)(-\Delta\varphi) + q\gamma D_x/\Delta^2(-\Delta T/T) \quad (36)$$
$$J_2 = (q\gamma D_x/\Delta^2)(-\Delta\varphi) + (q^2\gamma^2 D_x/\Delta^3 - \alpha T)\ (-\Delta T/\mathrm{T}). \quad (37)$$

In accordance with (36), the perlite growth rate is affected not only by the concentration thermodynamic force, but also by the temperature difference between the sample and the environment. Let us further consider the phase transformation of austenite under special conditions of steady growth of the pearlite colony, when it can be assumed that $\Delta T \approx$ const, $dT/dt \approx 0$. In this case, Eq. (37) takes the following form:

$$J_2 = (q\gamma D_x/\Delta^2)(-\Delta\varphi) + (q^2\gamma^2 D_x/\Delta^3 - \alpha T)\ (-\Delta T/\mathrm{T}) = 0. \quad (38)$$

For small ΔT, we can write approximately, as was done in [20]:

$$\Delta\varphi = \kappa\Delta T/T, \quad (39)$$

where k is the proportionality coefficient.

By analogy with the previously obtained solutions [21], we introduce the following notation:

$$\Delta_0 = \sqrt{\kappa q\gamma D_x/\alpha T}; \quad (40)$$

where $\Delta_l = q\gamma/k$ is the characteristic parameter of the system. (41)

Eq. (38) can now be represented in the form:

$$\Delta^3 - \Delta_0{}^2\Delta - \Delta_0{}^2\Delta_l = 0 \quad (42)$$

For $\Delta_l = 0$, as expected, the solution of Eq. (31) $\Delta = \Delta_0$. We obtain a well-known solution for the pearlite transformation of austenite [20]. In the real domain, there is one solution of Eq. (44). For small Δl (<0.5), the root Xk is in the region close to 1 ($Xk \to D_0$), with increasing Δ_l (in units of D_0), the root value increases. For large values of Δ_l, the root of Xk is approximately equal to

$$Xk \approx \sqrt[3]{\Delta l/\Delta_0}. \quad (43)$$

The between interplate distance of perlite for a stationary growth process is found from the formula:

$$S_0 = 2X\kappa \times \Delta_0 = 2\kappa Xk\sqrt{\Delta l D_x/\alpha T}. \qquad (44)$$

Using Eqs. (36) and (43), (44), we also find an improved expression for the perlite growth rate for an isothermal transformation

$$\frac{dX}{dt} = \frac{kDx}{S_0}\frac{\Delta T}{T}\left(1 + \frac{2\Delta l}{S_0}\right) \qquad (45)$$

The formula (45) is a more precise expression for determining the growth rate of perlite in the eutectoid transformation, than the expression obtained earlicr by the authors of [20].

We use the well-known dependence of the diffusion coefficient on temperature [17]:

$$D = A\ \exp.(-Q/RT),$$

где Q is the activation energy, ($Q \approx 134$ кJ/mol), and R is a constant ($R = 8314$ J/(mol·K)).

After substituting the known values of the steel parameters and taking into account that to 2.0, we find the calculated dependence of the perlite growth rate on the supercooling value of the alloy (**Figure 3**). In this figure, the dependence of the perlite growth rate on the supercooling value, calculated according to Zener's formula (1) [22, 23], is given for comparison.

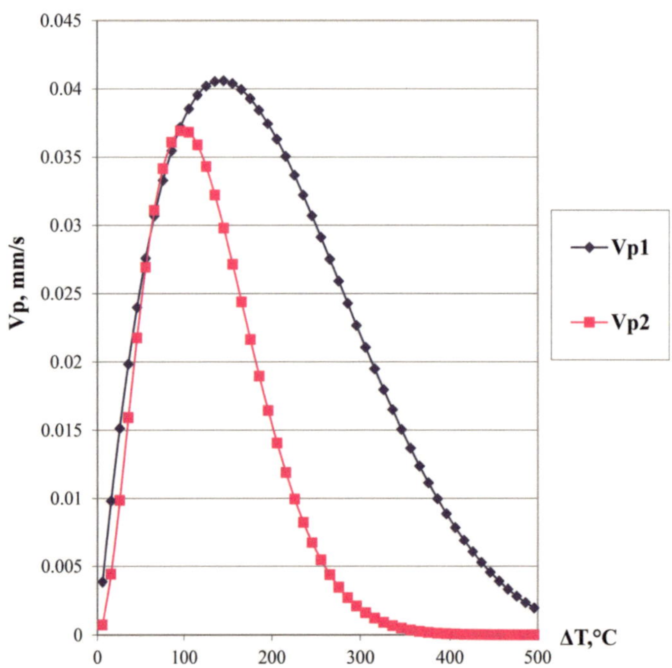

Figure 3.
Dependence of the perlite growth rate on the supercooling value, calculated from formula (47) of the present work (Vp1) *and Zener's formula (1) [22, 23]* (Vp2).

According to the constructed model, the perlite growth rate in the direction of the X axis has a maximum value at supercooling $\Delta T = 140.0°C$. The perlite growth rate calculated according to Zener's formula has a theoretical maximum value at overcooling $\Delta T = 96.0°C$. Consequently, the theoretical expressions (31) and (32) make it possible to describe with greater accuracy the maximum and the course of the experimental curve for the perlite formation rate presented in [3, 24] for high-purity eutectoid steel.

The expression for perlite growth rate obtained in this section has a significant value at supercooling of 300–400°C, thereby determining the possibility of perlite formation in this temperature range. Indeed, the formation of perlite in carbon steels in the temperature range 375–325°C was revealed in [24].

The calculated dependence of the between interplate distance of perlite by formula (46) on the magnitude of the supercooling of steel is shown in **Figure 4**. The same figure shows the experimental points from [24].

A fairly good agreement of the calculated dependence with the results of the latest experiments is observed, which indicates the adequacy of the proposed model.

5. Application of the positions of nonequilibrium thermodynamics to the analysis of the nondiffusion transformation of austenite

Martensite is the basis of hardened steel, so studying the mechanism and kinetics of its transformation is still of extreme interest for the theory and practice of heat treatment.

In the works of G.V. Kurdyumov and coworkers, the martensitic transformation is considered as the usual phase transformation in a one-component system, further complicated by the influence of a strong interatomic interaction, which leads to the development of significant stresses in the martensite crystal and matrix [25].

In accordance with the alternative mechanism, the martensitic transformation takes place by means of an instantaneous shift of atomic planes that does not require thermal activation and is not associated with thermodynamic transformation stimuli [1], [26]. In this case, the stress initiating the transformation is believed to be the stresses arising from the sharp cooling of the sample (quenching) [26].

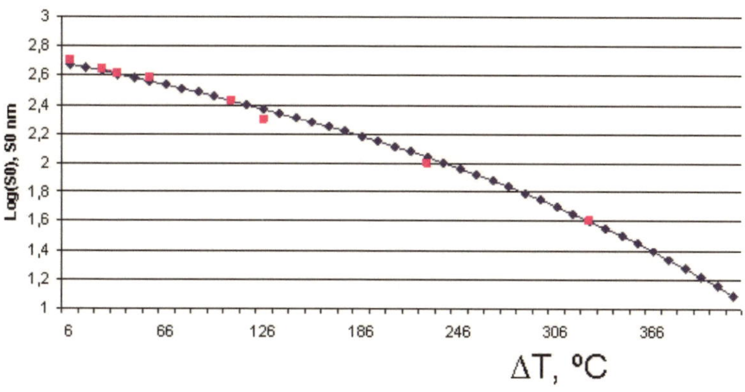

Figure 4.
The calculated dependence of the between interplate distance of perlite on the magnitude of the supercooling of steel (■ —experimental points from [24], p. 122, ■ —calculated points).

Considering the martensitic transformation as a thermally activated process, B.Ya. Lyubov used the equations of normal transformation obtained on the basis of the positions of nonequilibrium thermodynamics to describe his kinetics [3].

Changes in a complex or composite system under constant external conditions can be described as the process of increasing entropy. The rate of increase of entropy σ can be represented as the sum of the flux products and the corresponding forces for all transfer substrates in an amount of N [7–10]:

$$\sigma = \frac{dS}{dt} irrev = \sum_{\kappa=1}^{N} J_k X_k \text{ (k = 1, .., N)}, \tag{47}$$

In thc general case, the flows can be represented in the form (1).

The irreversible change in the entropy dS_{irrev} is equal to the sum of entropy changes in the system and the environment:

$$dS_{irrev} = dS + dS_e \tag{48}$$

Under isothermal conditions, when the released heat is absorbed by the environment and the temperature remains constant:

$$dS_e = - dQ/T, dQ = dU + PdV$$

$$dS_{irrev} = dS – (dUa + PdV)T^{-1} = (TdS – dU – PdV)T^{-1}. \tag{49}$$

Since $dU + PdV – TdS = dG$, and if we take into account the low compressibility of bodies in the condensed state and relatively small pressures, then.

$$\frac{dS}{dt} irrev = -T^{-1}\frac{dG}{dt} \approx - T^{-1}\frac{dF}{dt}, \tag{50}$$

where F is the free energy of the system.

The change in free energy in a system with a variable number of particles and internal stresses can be represented in the form [3], p. 142:

$$dF = dF\varepsilon + dFn = \sigma_{ik} d\varepsilon_{ik} + \varphi_l dn_l, \tag{51}$$

where $dF\varepsilon$ is the change in free energy in the system related to internal stresses, dFn is the change in the free energy in the system, determined by the variable number of particles of type l, σ_{ik} is the stress tensor, ε_{ik} is the strain tensor of the system, φ_l is the chemical potential of the lth element of the system, and n_l is the number of particles of the lst element of the system per unit volume, $l = 1, N$.

We now introduce some simplifying assumptions. First, for the nondiffusion transformation of austenite, only one kind of particles, the α-phase of iron n_α, will be taken into account. Approximately, this is also true for alloys of iron with close elements (nickel, chromium, cobalt). Of course, φ is some effective (averaged) chemical potential of the atoms of the alloy.

Secondly, we assume that the deformation of the system is a triaxial compression-expansion, and in the expression for $dF\varepsilon$, only the diagonal components of stress and strain tensors are taken into account:

$$\sigma_{ik} = \varepsilon_{ik} = 0, i \neq k. \sigma_{ii} = \sigma, \ \varepsilon_{ii} – \varepsilon \tag{52}$$

The change in internal energy can then be represented as:

$$dF = 3\sigma d\varepsilon + \varphi dn_{\alpha} \tag{53}$$

but the change in entropy:

$$\frac{dS}{dt}_{irrev} = -T^{-1}\left(3\sigma\frac{d\varepsilon_{\alpha}}{dt} + \varphi\frac{d\varepsilon_{\alpha}}{dt}\right). \tag{54}$$

Thus, in our system, in addition to the particle flow from the γ phase to the α-phase of $J_1 = dn_{\alpha}/dt$, we will also take into account the change in the strain of the sample with time $J_2 = d\varepsilon/dt$. These flows are related to the driving forces by the Eq. (1).

If, as charges of the process of nondiffusion transformation of austenite, the two quantities are the concentration of α-phase particles and the strain value, then, according to (1), the equations of motion take the form:

$$J_1 = L_{11}X_1 + L_{12}X_2 \tag{55}$$

$$J_2 = L_{21}X_1 + L_{22}X_2, \tag{56}$$

where $X_1 = \Delta\varphi$ is the thermodynamic force for iron, the change in the chemical potential at the transition of particles from the γ-phase to the α-phase, and $X_2 = \Delta\sigma$ is the change in the internal stress during the transition from the γ-phase to the α-phase.

The system of Eqs. (55) and (56) describes the contribution of stresses and deformations to the nondiffusion transformation of austenite. However, we do not yet know the coefficients of the equations in it. We now find expressions for the coefficients of the system of Eqs. (55) and (56). The coefficient L_{11} characterizes the normal transformation:

$$J_1 = \frac{dn_{\alpha}}{dt} = L_{11}(\varphi_{\gamma} - \varphi_{\alpha}) \tag{57}$$

In the normal kinetics of the phase transformation, the formation of the center (particle) of the α-phase occurs through separate (independent) acts of detachment of particles from the γ-phase and the attachment of atoms to the ferrite center. If we consider the process of formation of an α-phase close to the process of self-diffusion of iron in the γ-phase, then the coefficient L_{11} has the form [13]:

$$L_{11} = \frac{D_{\gamma}}{RT} \tag{58}$$

where D_{γ} is the self-diffusion coefficient of iron in the γ-phase (or the effective coefficient of self-diffusion in the γ-phase of the alloy), T is the transformation temperature, and R is the gas constant [27].

The self-diffusion coefficient of iron is taken in the usual notation [17]:

$$D_{\gamma} = D_0 e^{-\frac{U}{kT}} = 4.58 \cdot 10^{-4} exp(-252,000/\mathrm{RT}) \tag{59}$$

where D_0 is a multiplier and U is the activation energy of diffusion.

The coefficient L_{22} characterizes the direct relationship:

$$J_2 = \frac{d\varepsilon_{\alpha}}{dt} = L_{22}(\sigma_{\gamma} - \sigma_{\alpha}). \tag{60}$$

Let $\sigma_{\gamma} = 0$. Let us take into account that for triaxial compression stretching [28]: where σ_{α} is the stress in the α phase and σ_{γ} is the stress in the γ phase.

Let $\sigma_\gamma = 0$. Let us take into account that for triaxial compression-stretching [28]:

$$\sigma_\alpha = \frac{\mathrm{E}}{3(1\text{-}2\mu)}\frac{\Delta V}{V} = \frac{\mathrm{E}}{1\text{-}2\mu}\varepsilon_\alpha, \tag{61}$$

where E is the modulus of elasticity of steel ($\sim 2.17 \cdot 10^5$ MPa) and μ is the Poisson ratio (~ 0.26).

Then, expression (60) can be transformed as follows:

$$\frac{d\varepsilon_\alpha}{dt} = L_{22}\sigma_\alpha = L_{22}\frac{\mathrm{E}}{1\text{-}2\mu}\varepsilon_\alpha = \frac{\mathrm{v}}{L}\varepsilon, \tag{62}$$

where the following values are entered:

v is the propagation velocity of the microdeformation in sample (~1000 m/c) [3] and L is the characteristic distance over which the microdeformation of the shear is propagated (the size of the martensitic strips or plates). At the initial stage of the formation of the shear structure, it has a magnitude of the order of the diameter of the austenite grain (~ 100 μm), and then decreases with decreasing temperature [1].

From Eq. (62), we find that the coefficient L_{22} is equal to:

$$L_{22} = \frac{\mathrm{v}(1\text{-}2\mu)}{LE}. \tag{63}$$

The cross-coefficients $L_{12} = L_{21}$ for a nonequilibrium thermodynamic system are found with sufficient accuracy by the formulas proposed in [5]:

$$L_{12} = \sqrt{L_{11}L_{22}} = \sqrt{\frac{D_\gamma}{RT}\frac{\mathrm{v}(1\text{-}2\mu)}{LE}} \tag{64}$$

Thus, we obtained simple differential equations for a nonequilibrium thermodynamic system describing the nondiffusion transformation of austenite taking into account the influence of internal stresses.

Let us write the equations of motion of our system in the form:

$$\frac{dn_\alpha}{dt} = L_{11}\Delta\varphi + L_{12}\sigma_\gamma - L_{12}\sigma_\alpha. \tag{65}$$

$$\frac{d\varepsilon_\alpha}{dt} = L_{21}\Delta\varphi + L_{22}\sigma_\gamma - L_{22}\sigma_\alpha. \tag{66}$$

We first transform Eq. (66) taking into account expression (62). We have:

$$\frac{d\varepsilon_\alpha}{dt} + v\varepsilon_\alpha/L = L_{21}\Delta\varphi + L_{22}\sigma_\gamma, \tag{67}$$

where ε_α is the magnitude of deformations of the α-phase. The differential Eq. (66) with constant coefficients (temperature) has a solution:

$$\varepsilon_\alpha = \frac{(\mathrm{L}_{21}\Delta\varphi + L_{22}\sigma_\gamma)\ (1\text{-}2\mu)}{L_{22}E}\left(1 - e^{-\frac{vt}{L}}\right). \tag{68}$$

This kinetic equation describes the change in the magnitude of the deformation of the α-phase in time. At $t = 0$, $\varepsilon_\alpha = 0$. When the time is counted, a fast ($\sim 10^{-6}$ s) process of transition to deformation occurs:

$$\varepsilon_\alpha = \varepsilon_\gamma + \Delta\varepsilon_\alpha = \varepsilon_\gamma + \frac{L_{21}\Delta\varphi(1\text{-}2\mu)}{L_{22}E}. \tag{69}$$

Eq. (69) shows that the residual deformation of the α-phase after the transient process consists of the deformation of the austenite ε_γ and the additional deformation $\Delta\varepsilon_\alpha$. This additional deformation determines the change in the volume of the sample as $\gamma \to \alpha$-transformation:

$$\frac{\Delta V_{\gamma\to\alpha}}{V_\gamma} = 3n_\alpha\Delta\varepsilon_\alpha. \tag{70}$$

Then, substituting expression (53) into Eq. (51.1), we find:

$$\begin{aligned}\frac{dn_\alpha}{dt} &= L_{11}\Delta\varphi + L_{12}\sigma_\gamma - \frac{L_{12}(L_{21}\Delta\varphi + L_{22}\sigma_\gamma)}{L_{22}}\left(1 - e^{-\frac{v}{L}t}\right) = L_{11}\Delta\varphi + L_{12}\sigma_\gamma - \\ &(L_{11}\Delta\varphi + L_{12}\sigma_\gamma)\left(1 - e^{=\frac{v}{L}t}\right) = (L_{11}\Delta\varphi + L_{12}\sigma_\gamma)e^{-\frac{v}{L}t}.\end{aligned} \tag{71}$$

It can be concluded from expression (71) that the growth rate of α-phase particles depends on the stresses in the γ-phase. The greater the value of tensile stresses in the γ phase, the higher the growth rate of ferrite particles. The rate of growth of the α-phase particles at a constant temperature very rapidly (exponentially) decreases in time, determining the incompleteness of the transformation.

Integration of Eq. (71) with time-independent coefficients L_{11} and L_{12} allows us to obtain the kinetic equation for n_α:

$$n_{\alpha=}\frac{(L_{11}\Delta\varphi + L_{12}\sigma_\gamma)L}{v}\left(1 - e^{-\frac{v}{L}t}\right). \tag{72}$$

In accordance with Eq. (72), the amount of α-phase formed depends not only on the thermodynamic force $\Delta\varphi$, but also on the magnitude of the stresses in the γ-phase.

6. Scheme of the nondiffusion transformation of austenite based on the constructed model.

Before discussing the equations obtained, we introduce some more useful relations characterizing the $\gamma \to \alpha$ transformation. With the $\gamma \to \alpha$ transformation, the effective atomic volume of the iron lattice changes in the sample under consideration, characterized by $\Delta V_{\gamma\to\alpha}$ and the relative volume change $\frac{\Delta V_{\gamma\to\alpha}}{V_\gamma}$. According to the data of [3]:

$$\Delta V_{\gamma\to\alpha} = 0.268 - 1.62^{*}10^{-4}T,\ \text{sm}^3/\text{mol} \tag{73}$$

We will assume that with the formation of the α-phase, the relative change in volume is determined by the additional deformation: $\frac{\Delta V_{\gamma\to\alpha}}{V_\gamma} = 3\Delta\varepsilon_\alpha$, and the compressive stress arising in the α-phase has the value.

$$\sigma_\alpha = \frac{\text{E}}{1\text{-}2\mu}\Delta\varepsilon_\alpha. \tag{74}$$

When the alloy sample is cooled by ΔT, a deformation occurs in its surface layer: $\varepsilon_\gamma \sim \alpha\Delta T$ and the tensile stress σ_γ corresponding to this deformation:

$$\sigma_\gamma = \frac{E}{1\text{-}2\mu}\alpha\Delta T. \tag{75}$$

Comparing the values of thermodynamic forces among themselves, it is possible to classify the types of nondiffusion transformation according to the kinetic criterion. As shown in [1], p. 208, for small deviations of the system from equilibrium, the growth of crystals is more likely, controlled by self-diffusion, at large–cooperative growth. The same phase transition in a single-component system under different external conditions can take place with an independent (or slightly dependent) temperature growth rate (martensitic kinetics) and with a rate that exponentially depends on the temperature at an activation energy close to the activation energy of self-diffusion (normal kinetics). The parameter characterizing the deviation of the system from equilibrium is the supercooling of the alloy $\Delta T = Ac_3 - T$, where Ac_3 is the temperature of the end $\alpha \to \gamma$ of the conversion upon heating, and T is the transformation temperature. The transformation scheme for the constructed model is shown in **Figure 5**.

Ac_1 is the temperature of the beginning of $\alpha \to \gamma$ transformation when the alloy is heated and Mni is the temperature of the onset of the formation of isothermal martensite upon supercooling of the alloy. Mn is the temperature of the onset of athermal martensite formation upon supercooling of the alloy. Mk is the temperature of the end of martensite formation upon supercooling of the alloy.

Thus, for small

$$\Delta T : L_{11}\Delta\varphi > L_{12}\sigma_\alpha > L_{12}\sigma_\gamma, \tag{76}$$

then the growth of α-phase crystals is determined by self-diffusion by the normal mechanism. However, as follows from Eq. (72), in this case too, the contribution of deformations (and stresses) to the conversion kinetics is very significant. In order that the condition (76) is satisfied, it is necessary that the stress level in the γ- and α-phases be small; for the α-phase, this is possible only in the case of relaxation of internal stresses in the alloy at high temperature by the mechanism of recrystallization.

With increasing supercooling of the alloy, the thermodynamic stimulus and the rate of normal transformation increase.

For a larger

$$\Delta T : L_{11}\Delta\varphi \sim L_{12}\sigma_\alpha > L_{12}\sigma_\gamma \tag{77}$$

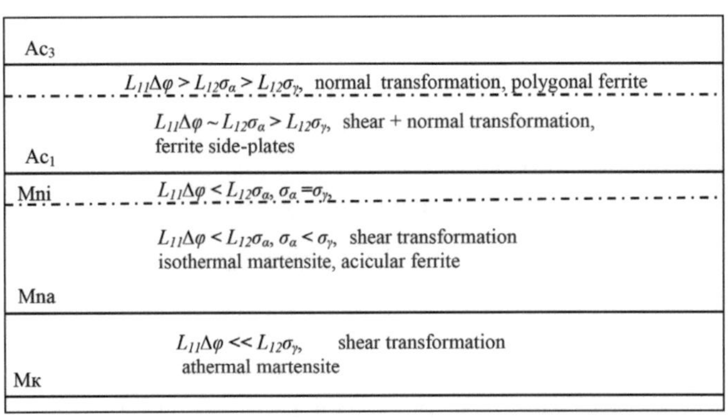

Figure 5.
Scheme of nondiffusion transformations from the constructed model.

The existing thermal stresses in the γ phase (75) contribute to the formation of the α-phase by the shear mechanism, and the stresses arising in the α-phase compensate thermal stresses in the γ-phase. With a certain amount of α-phase, the stress equals $\sigma_\alpha = \sigma_\gamma$ arises and the further formation of the α-phase occurs according to the normal mechanism with the relaxation of the arising stresses by recrystallization. Consequently, the condition (77) corresponds to the transformation of the γ-phase by a mixed mechanism, and also to the formation of a ferrite side plates (Widmanstätten), followed by the release of the α-phase by the normal mechanism [1].

With a certain supercooling of ΔTi, stress compensation occurs only when the γ-phase is completely transformed into ferrite by a shearing mechanism. In this case:

$$L_{11}\Delta\varphi \sim L_{12}\sigma_\alpha,\ \sigma_\alpha = \sigma_\gamma$$

The temperature corresponding to this supercooling is the starting point for the formation of the isothermal martensite Mni (**Figure** 5). Below the point Mni, the formation of the α-phase occurs by a shearing mechanism. However, the normal component of the process still has a significant value, affecting the morphology of the resulting precipitates. When supercooling a greater ΔTi, $L_{11}\Delta\varphi < L_{12}\sigma_\alpha$, $\sigma_\alpha < \sigma_\gamma$

At temperatures below Mni, isothermal martensite or acicular ferrite is formed with a "reticular" or acicular morphology of precipitates. Finally, for large ΔT (below Mna):

$$L_{11}\Delta\varphi \ll L_{12}\sigma_\gamma, \quad (78)$$

Inequality (78) determines the condition for the formation of "athermal" martensite, when the normal component does not affect the formation of the shear structure. The main effect on the rate of the $\gamma \to \alpha$ transformation, in accordance with expression (71), is due to thermal stresses in the γ phase. Thus, the constructed model of the nondiffusion austenite transformations allows us to consider the normal and martensitic transformations, as limiting cases.

7. Conclusions

Based on the possibility of dynamic equilibrium, expressions are found for calculating the cross-kinetic coefficients of a thermodynamic system consisting of two and three components. The values of the thermodynamic force for diffusion of carbon, kinetic coefficients and flows of a thermodynamic system describing the kinetics of carbide precipitation during the tempering of chromium steel are calculated. It has been established that the values of iron and chromium fluxes increase substantially due to the cross ratios and the significant magnitude of the thermodynamic force ($-\Delta\mu_C$).

Analysis of the eutectoid transformation of austenite using the relations of nonequilibrium thermodynamics allowed us to generalize the equations of motion of the system obtained earlier by the authors of [20] and to find more accurate theoretical expressions for the perlite growth rate and its between interplate distance on the magnitude of the supercooling of steel. According to the constructed model, the perlite growth rate in the direction of the X axis has a maximum value at supercooling $\Delta T = 140.0°C$. The perlite growth rate calculated according to Zener's formula has a theoretical maximum value at overcooling $\Delta T = 96.0°C$. Consequently, the theoretical expressions (31) and (32) make it possible to describe with greater accuracy the maximum and the course of the experimental curve for the perlite formation for high-purity eutectoid steel.

The application of nonequilibrium thermodynamics to the analysis of the nondiffusion transformation of austenite made it possible to obtain a system of equations for the thermodynamic system and to generalize the results obtained earlier by B.Ya. Lyubov the equations for a normal transformation. The theoretical expression for the growth rate of the α-phase, obtained in this paper, takes into account the influence of stresses on the process of austenite transformation. It is shown that the rate of growth of α-phase particles at a constant temperature very rapidly (exponentially) decreases in time, determining the incompleteness of the transformation. According to the constructed model, a scheme of nondiffusion austenite transformations was developed, including normal and martensitic transformations, as limiting cases.

Thus, the use of the principles of nonequilibrium thermodynamics makes it possible to obtain completely new results in the analysis of phase transformations in iron-carbon alloys.

Author details

Bobyr Sergiy Volodimyrovych
Iron and Steel Institute of Z.I. Nekrasov, National Academy of Sciences of Ukraine, Dnipro, Ukraine

*Address all correspondence to: svbobyr07@gmail.com

References

[1] Umanskiy YS, Skakov YA. Fizika metallov. M.: Atomizdat. Moscow; 1978

[2] Kurdyumov GV, Utevskiy LM, Entin RI. Prevrascheniya v zheleze i stali. M Nauka. Moscow; 1977

[3] Ya Lyubov B. Kineticheskaya teoriya fazovyih prevrascheniy. M.: Metallurgiya; 1969. 264 p

[4] Bobyr SV, Bolshakov VI. Uspehi fiziki metallov. Modeli I Harakteristiki Preryivistogo Prevrascheniya Austenita V Zhelezouglerodistyih Splavah. 2014; **T.15**:123

[5] Bobyir SV. Metallofizika i Noveishie Tekhnologii. 2013;**35**(2):35

[6] Krishtal MA, Zhukov AA, Snezhnoy RL, Titenskiy EG. Termodinamika, fizicheskaya kinetika strukturoobrazovaniya i svoystva chuguna i stali. Vyip. 4. Moskva: Metallurgiya; 1971

[7] Prigozhin I. Vvedenie v termodinamiku neobratimyih protsessov. Introduction to Thermodynamics of Irreversible Processes. Moskva: IL; 1960

[8] de Groot S, Mazur P. Neravnovesnaya termodinamika. Moskva: Mir; 1964

[9] Onsager L. Physics Review. 1931;**37**: 405

[10] Veynik AI. Termodinamika. Minsk: Vyisheyshaya shkola; 1968

[11] Zhukov AA, Snezhnoy RL. Diffuzionnyie protsessyi v metallah. Kiev: Naukova dumka; 1966

[12] Baranov AA, Bunin KP. Diffuzionnyie protsessyi v metallah. Kiev: Naukova dumka; 1966

[13] Bokshteyn BS. Diffuziya v metallah. Moskva: Metallurgiya; 1978

[14] Bokshteyn SZ. Struktura i mehanicheskie svoystva legirovannoy stali. M.: Metallurgizdat. Moscow; 1954

[15] Silman GI. Zhurnal fizicheskoy himii. Metodika Rascheta Diagramm Sostoyaniya Troynyih Sistem S Ispolzovaniem Koeffitsientov Mezhfaznogo Raspredeleniya Elementov Ch.1. Dvuhfaznoe Ravnovesie. 1983;**57**(2):307-313

[16] Silman GI. Uglerodnyie ekvivalentyi elementov v chugunah. Metallovedenie I Termicheskaya Obrabotka Metallov (MiTOM). 2002;**1**:S.26-29

[17] Larikov LN, Isaychev VI. Diffuziya v metallah i splavah. Spravochnik. Kiev: Naukova dumka; 1987

[18] Bobyr SV. Neravnovesnyiy termodinamicheskiy analiz protsessa obrazovaniya karbidov v hromistoy stali pri otpuske Metallofizika i noveyshie tehnologii. 2014;**T.36**:1385

[19] Bobyr SV, Bolshakov VI, Sharfnadel SA. Material Science. Nonequilibrium Phase Transformations. Sofia, Bulgaria Is. 4; 2016

[20] Bolshakov VI, Bobyr SV. MiTOM. Kineticheskie parametry obrazovaniya perlita v zhelezouglerodistykh splavakh [Kinetic parameters of pearlite formation in iron-carbon alloys]. 2004;(8)

[21] Bobyr SV, Bolshakov VI. Novi materiali i tehnologiyi v metalurgiyi ta mashinobuduvanni. 2012;**1**:21

[22] Meyl RF, Hagel UK. Uspehi fiziki metallov. Austenitno-perlitnoe Prevraschenie. 1960;**III**:86-156. M: GosNTI

[23] Zener C. Transactions of AIME. 1946;**V**:167

[24] Schastlivtsev VM, Mirzaev DA, Yakovleva IL. i dr. Perlit v uglerodistyih stalyah. Ekaterinburg: UrO RAN; 2006

[25] Kurdyumov GV. Yavleniya zakalki i otpuska stali. M.: Metallurgizdat; 1960

[26] Oshkaderov SP. Uspehi fiziki metallov; 2011;**T.12**:269

[27] Smithells CJ. Metalli—M Metallurgia; 1980

[28] Belous MV, Braun MP. Fizika metallov—K.: Vischa shkola; 1986

Chapter 5

Variational Principle for Nonequilibrium Steady States Tested by Molecular Dynamics Simulation of Model Liquid Crystal Systems

Sten Sarman, Yonglei Wang and Aatto Laaksonen

Abstract

The purpose of the work presented in this chapter is to test a recently proven variational principle according to which the irreversible energy dissipation rate is minimal in the linear regime of a nonequilibrium steady state. This test is carried out by performing molecular dynamics simulations of liquid crystals subject to velocity gradients and temperature gradients. Since the energy dissipation rate varies with the orientation of the director of the liquid crystal relative to these gradients and is minimal at certain orientations, this is a stringent test of the variational principle. More particularly, a nematic liquid crystal model based on the Gay-Berne potential, which can be regarded as a Lennard-Jones fluid generalized to elliptical molecular cores, is studied under planar Couette flow, planar elongational flow, and under a temperature gradient. It is found that the director of a nematic liquid crystal consisting of rod-like molecules lies in the vorticity plane at an angle of about 20° to the stream lines in the planar Couette flow. In the elongational flow, it is parallel to the elongation direction, and it is perpendicular to the temperature gradient in a heat flow. These orientations are the ones where the irreversible energy dissipation rate is minimal, so that the variational principle is fulfilled in these three cases.

Keywords: liquid crystals, nonequilibrium molecular dynamics simulation, shear flow, elongational flow, heat conduction, alignment phenomena, minimal energy dissipation rate

1. Introduction

For a system in thermodynamic equilibrium, there is a variational principle according to which the free energy is minimal, that is, the Helmholtz free energy when the volume, temperature, and number of particles are constant, Gibbs free energy when the pressure, temperature, and number of particles are constant, etc. On the other hand, for systems driven away from equilibrium by an external dissipative field such as a velocity gradient, temperature gradient, chemical potential gradient or an electrical potential gradient doing irreversible work that is converted to heat, there has not been any variational principle to date. However, a theorem originally proposed by Ilya Prigogine stating that a quantity, known as the irreversible energy dissipation rate, $\dot{w}_{irr}$, is minimal in the linear regime of a nonequilibrium steady state has recently been proven [1]. This quantity is defined as the irreversible work per unit time and unit volume that is done by a dissipative external field on the system [2]. Thus, there is a variational principle for nonequilibrium steady states.

This theorem is not only of basic scientific interest but also of technological and practical interest since shear fields, temperature gradients, concentration gradients, or chemical potential gradients and electrical potential gradients are common examples of external dissipative fields that are ubiquitous in industrial applications and in everyday life. For example, in a lubricated bearing, a planar Couette flow arises in the lubricant in the narrow space between two surfaces rotating at different angular velocities, and $\dot{w}_{irr}$ is equal to the product of the shear rate and the shear stress. Another example is the heat flow between a hot region and a cold region such as the inside and the outside of a building. Then, $\dot{w}_{irr}$ is equal to the product of the heat flow and the temperature gradient. Still another example is an electric heating element where an electric potential difference or voltage drives an electric current, and $\dot{w}_{irr}$ is equal to the product of the voltage and the current. Finally, chemical potential gradients arise when various substances are mixed and they begin to diffuse, and $\dot{w}_{irr}$ is equal to the product of the chemical potential gradients and the matter currents.

One way of testing this principle is to perform molecular dynamics simulations of microscopic model systems, but then it is hard to find a suitable model system that is easy to analyze. However, liquid crystals are particularly interesting for this purpose because the transport properties and thereby $\dot{w}_{irr}$ depend on the orientation relative to the streamlines or the temperature gradient, and at certain orientations, $\dot{w}_{irr}$ is minimal. Thus, it can be determined whether these orientations actually are attained by the liquid crystal. Moreover, it is possible to orient the liquid crystal in an experimental measurement by applying an electric or magnetic field and in molecular dynamics simulations by applying a constraint torque. This means that $\dot{w}_{irr}$ can be measured or calculated as a function of the orientation relative to the dissipative field.

The simplest kind of liquid crystal is the nematic liquid crystal [3, 4]. It consists of rod-like or plate-like molecules oriented in a certain direction—the director—but there is no translational order, see **Figure 1**. A nematic liquid crystal cannot support shear stresses, so it is by definition a liquid, but it can support torques, which is the basis for various orientation phenomena relative to external fields. A special case of a nematic liquid crystal is the cholesteric liquid crystal, where the director rotates in space around an axis perpendicular to itself—the cholesteric axis or the optical axis. The spatial rotation period or the pitch is of the order of 1 μm or about 500 molecular diameters. A cholesteric liquid crystal is different from its mirror image, and it is formed by chiral molecules.

There is some theoretical and experimental evidence indicating that the director comes to rest in an orientation where the irreversible energy dissipation rate is minimal in accordance with the variational principle. More specifically, such orientation phenomena have been observed in simulations of shear flow or planar Couette flow [5, 6], in experimental measurements of the viscosity [7] in this flow geometry, and in simulations of planar elongational flow [8]. In the latter case, it is actually possible to prove that the energy dissipation rate must be either minimal or maximal in a steady state in the linear or Newtonian regime by using the linear phenomenological relations between the velocity gradient and the shear stress.

In the case of a nematic liquid crystal subject to a temperature gradient, there are quite a few early experimental works [9–14] that might imply that the director of a

Figure 1.
A nematic phase of the Gay-Berne fluid undergoing planar Couette flow. The velocity gradient is directed in the vertical direction and the streamlines are directed in the horizontal direction. Note that the director forms an angle with streamlines of about 18°.

liquid crystal consisting of rod-like molecules orients perpendicularly to this gradient. This means that the heat flow is minimized, since the heat conductivity is minimal in this orientation. Unfortunately, the results of these works are not wholly conclusive because the underlying experiments are very hard to carry out. On the other hand, molecular dynamics simulation of nematic phases of calamitic and discotic soft ellipsoids [15–17] clearly show that the directors orient perpendicularly and parallel, respectively, to the temperature gradient, so that the heat flow and thereby $\dot{w}_{irr}$ are minimized. However, one system, where the director definitely orients perpendicularly to the temperature gradient, is the cholesteric liquid crystal, where the cholesteric axis orients parallel to the temperature gradient, so that the director becomes perpendicular to this gradient, and the heat flow is minimized [3, 4, 18–20], which is in agreement with the variational principle.

This chapter is organized in the following way: in Section 2, the basic theory is outlined, and in Sections 3, 4, and 5, molecular dynamics simulation results and experimental measurements on the director orientation and the irreversible energy dissipation rate are presented and discussed for shear flow or planar Couette flow, planar elongational flow and heat conduction, respectively. In Section 6, the effects of the thermostat are discussed, and finally in Section 7, there is a conclusion. Some background theory is given in the Appendices.

2. Basic theory

2.1 Order parameter, director, and director angular velocity

In order to describe transport properties of a liquid crystal, we must first define the order parameter, the director, and the director angular velocity. In an axially symmetric system such as a nematic or a smectic A liquid crystal, the order parameter, S, is given by the largest eigenvalue of the order tensor,

$$\mathbf{Q} = \frac{3}{2}\left(\frac{1}{N}\sum_{i=1}^{N}\hat{\mathbf{u}}_i\hat{\mathbf{u}}_i - \frac{1}{3}\mathbf{1}\right), \quad (1)$$

where N is the number of particles, and $\{\hat{\mathbf{u}}_i; 1 \leq i \leq N\}$ is some characteristic vector of the molecule; in the case of bodies of revolution, it can be taken to be parallel to the axis of revolution but in a more realistic all atom model some other vector in the molecule has to be defined as $\hat{\mathbf{u}}_i$, and $\mathbf{1}$ is the unit second rank tensor. When the molecules are perfectly aligned in the same direction, the order parameter is equal to unity, and when the orientation is completely random, it is equal to zero. The eigenvector corresponding to the order parameter is defined as the director, $\mathbf{n}$, and it is a measure of the average orientation of the molecules in the system. The order tensor can also be expressed as

$$\mathbf{Q} = \frac{3}{2} S \left(\mathbf{nn} - \frac{1}{3} \mathbf{1} \right). \tag{2}$$

The director angular velocity is given by $\mathbf{\Omega} = \mathbf{n} \times \dot{\mathbf{n}}$. In a macroscopic system, the order tensor and the order parameter are functions of the position in space, but in a small system such as a simulation cell with dimensions of the order of some ten molecular lengths, there is only one director and one order parameter for the whole system.

2.2 Director constraint algorithm

Since the molecules studied in the work presented in this chapter are modeled by the Gay-Berne potential, which can be regarded as a Lennard-Jones potential generalized to elliptical molecular cores, see Appendix 2, they are rigid bodies. Therefore, the Euler equations are applied in angular space. Moreover, since the purpose often is to find the stable orientations of the director relative to an external dissipative field, it is interesting to calculate the torque exerted on the liquid crystal, when the director attains various fixed angles relative to this field. This can be done by adding Gaussian constraints to the Euler equations [21],

$$I \dot{\boldsymbol{\omega}}_i = \boldsymbol{\Gamma}_i + \lambda_x \frac{\partial \Omega_x}{\partial \boldsymbol{\omega}_i} + \lambda_y \frac{\partial \Omega_y}{\partial \boldsymbol{\omega}_i}, \tag{3}$$

where I is the moment of inertia around the axes perpendicular to the axis of revolution, $\boldsymbol{\omega}_i$ is the angular velocity of molecule i, $\boldsymbol{\Gamma}_i$ is the torque exerted on molecule i by the other molecules, Ω_x and Ω_y are the x- and y-components of the director angular velocity, and λ_x and λ_y are Gaussian constraint multipliers keeping the x- and y-components of the director angular acceleration equal to zero. These multipliers are determined in such a way that the director angular acceleration becomes a constant of motion. Then if the initial director angular acceleration and angular velocity are equal to zero, the director will remain fixed in space for all subsequent times and the time averages of the constraint multipliers will be equal to the torque exerted on the director by the external field. Finally, note that the difference between the director angular velocity, $\mathbf{\Omega}$, and the molecular angular velocities $\{\boldsymbol{\omega}_i; 1 \leq i \leq N\}$; the director angular velocity can be regarded as the angular velocity of the average orientation of the molecules. If the director angular velocity is constrained to be zero by applying Eq. (3), the molecular angular velocities are still nonzero and the right hand side of Eq. (3) is nonzero. The Gaussian constraint simply forces the molecules to rotate in such a way that average orientation stays the same.

3. Shear flow

3.1 The SLLOD equations of motion for shear flow

In order to study shear flow and to calculate the viscosity and director alignment angles relative to the streamlines, it is convenient to apply the SLLOD equations of motion [22]. The name SLLOD stems from the similarity to the Dolls equation of motion derived from the Dolls tensor Hamiltonian. They are synthetic equations of motion that can be used to calculate the viscosity in the linear regime. On the other hand, the idea behind the SLLOD equations of motion is very simple: The velocity of the molecules is divided into the streaming velocity and the thermal velocity. The thermal velocity is related to the temperature, and the streaming velocity is the macroscopic external velocity. The SLLOD equations of motion are an exact description of adiabatic planar Couette flow and a very good approximation of shear flow at constant temperature both in the linear and nonlinear regime. The SLLOD equations are expressed in the following way:

$$\dot{\mathbf{r}}_i = \frac{\mathbf{p}_i}{m} + \gamma r_{zi}\mathbf{e}_x \tag{4a}$$

and

$$\dot{\mathbf{p}}_i = \mathbf{F}_i - \gamma p_{zi}\mathbf{e}_x - \alpha \mathbf{p}_i, \tag{4b}$$

where $\mathbf{r}_i$ and $\mathbf{p}_i$ are the position and peculiar momentum, that is, the momentum relative to the streaming velocity, of molecule i, m is the molecular mass, $\gamma = \partial u_x/\partial z$ is the shear rate, that is, there is a streaming velocity u_x in the x-direction varying linearly in the z-direction, see **Figure 2**, $\mathbf{e}_x$ is the unit vector in the x-direction, $\mathbf{F}_i$ is the force exerted on molecule i by the other molecules, and α is a thermostatting multiplier given by the constraint that the linear peculiar kinetic energy is a constant of motion,

$$\alpha = \frac{\sum_{i=1}^{N}\left[\mathbf{F}_i \cdot \mathbf{p}_i - \gamma p_{ix}p_{iz}\right]}{\sum_{i=1}^{N}\mathbf{p}_i^2}. \tag{5}$$

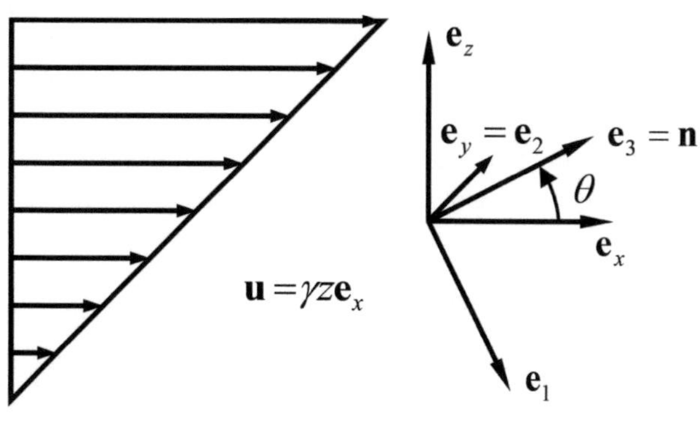

Figure 2.
Planar Couette flow or shear flow arises when there is a streaming velocity in the x-direction, varying linearly in the z-direction, $\mathbf{u} = \gamma z\mathbf{e}_x$, where $\gamma = \partial u_x/\partial z$ is the shear rate or velocity gradient. The expression for the relation between the velocity gradient and the pressure tensor becomes simpler by using a director-based coordinate system $(\mathbf{e}_1, \mathbf{e}_2, \mathbf{e}_3)$, where the director $\mathbf{n}$ points in the $\mathbf{e}_3$-direction, obtained by rotating the ordinary laboratory-based coordinate system $(\mathbf{e}_1, \mathbf{e}_2, \mathbf{e}_3)$ with an angle Θ around the $\mathbf{e}_y = \mathbf{e}_2$-axis. Reproduced from Ref. [6] with the permission of AIP Publishing.

This expression is obtained by applying Gauss's principle of least constraint [22]. This principle is essentially the same as the Lagrange's method for handling constraints. However, Gauss's principle is more general in that it in addition to constraints involving the molecular coordinates also allows handling of some constraints involving the molecular velocities. This is very useful because it makes it possible to keep the kinetic energy constant whereby the temperature also will be constant. It is possible to show that the ensemble averages of the phase functions and the time correlation functions are essentially canonical when a Gaussian thermostat is applied.

3.2 Shear flow of nematic liquid crystals

In a nematic liquid crystal undergoing shear flow, the alignment angle, θ, between the director and the streamlines is determined by a mechanical stability criterion, namely, that the antisymmetric pressure must be zero when no external torques act on the system, that is, that the torques exerted by the vorticity and the strain rate cancel out. This makes it possible to derive a relationship between the alignment angle and the viscosity coefficients in the Newtonian regime by using the linear relation between the pressure tensor and the strain rate, see Refs. [3, 4, 23] and Appendix 1,

$$\langle p_2^a \rangle = -\tilde{\gamma}_1 \frac{\gamma}{4} - \tilde{\gamma}_2 \frac{\gamma}{4} \cos 2\theta = 0, \tag{6}$$

where $\tilde{\gamma}_1$ is the twist viscosity, $\tilde{\gamma}_2$ is the cross coupling coefficient between the antisymmetric pressure and the strain rate, and $\langle p_2^a \rangle$ is the antisymmetric pressure in the vorticity direction perpendicular to the streamlines and perpendicular to the velocity gradient. The angular brackets denote that the pressure tensor is the ensemble average of a phase function. Then, if $\langle p_2^a \rangle$ is equal to zero, we obtain

$$\cos 2\theta_0 = -\tilde{\gamma}_1/\tilde{\gamma}_2, \tag{7}$$

for the preferred alignment angle, θ_0, provided that the ratio $|\tilde{\gamma}_1/\tilde{\gamma}_2|$ is less than one. Then the liquid crystal is said to be flow stable. This condition is fulfilled in many liquid crystals, and θ_0 is between 10 and 20° both in real systems and in simplified coarse grained model systems such as the soft ellipsoid fluid, see Refs. [5–7], **Figure 1**, and Appendix 2. Note, however, that for some systems, often near the nematic-smectic A phase transition, the ratio $|\tilde{\gamma}_1/\tilde{\gamma}_2|$ is greater than one. This means that there is no orientation angle where the antisymmetric pressure is zero, so that no steady state is attained. Then the liquid crystal is said to be flow unstable and the director will rotate forever [3, 4, 23, 24].

The connection with the variational principle can be made by using the fact that there is an algebraic expression for the irreversible energy dissipation rate, $\dot{w}_{irr}$, of a flow stable nematic liquid crystal, given by the dyadic product of the symmetric traceless pressure, $\overline{\mathbf{P}}$, and the traceless strain rate, $\overline{\nabla \mathbf{u}}$,

$$\dot{w}_{irr} = -\overline{\mathbf{P}} : \overline{\nabla \mathbf{u}} = \left(\eta + \frac{\tilde{\eta}_1}{6} + \frac{\tilde{\eta}_3}{2} \sin^2 2\theta + \frac{\tilde{\eta}_2}{2} \cos 2\theta \right) \gamma^2, \tag{8}$$

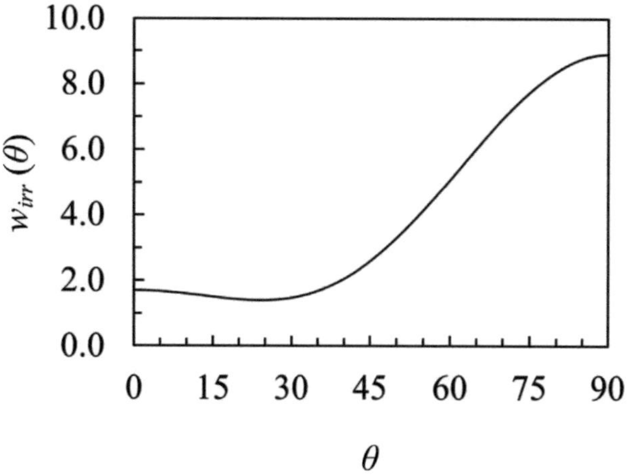

Figure 3.
The irreversible energy dissipation rate, $\dot{w}_{irr}$, Eq. (8), due to the strain rate of a nematic liquid crystal phase of calamitic soft ellipsoids as a function of the director alignment angle, Θ, is obtained by using the Gaussian constraint algorithm (3) to fix the director at various angles relative to the streamlines. The preferred alignment angle attained when no constraints are applied is equal to about 20° which is close to the minimum of $\dot{w}_{irr}$. Reproduced from Ref. [6] with the permission of AIP Publishing.

where the definitions of the viscosity coefficients, η, $\tilde{\eta}_1$, $\tilde{\eta}_2$, and $\tilde{\eta}_3$, and the derivation are given in Appendix 1. If the values of the various viscosity coefficients are inserted, it is found that the functional dependence of $\dot{w}_{irr}$ on θ is similar to that given in **Figure 3**. This function (8) has been obtained by shear flow simulations applying the SLLOD equations of motion [22] to a nematic phase of calamitic soft ellipsoids, see Refs. [5, 6]. The energy dissipation rate is low close to the preferred alignment angle and high when the director is perpendicular to the streamlines and parallel to the velocity gradient. Then, if we study the distribution of the director, we find that it is centered close to the minimum of $\dot{w}_{irr}$. This minimum has also been observed in simulations of shearing nematic phases of discotic soft ellipsoids [5] and when experimentally measured, viscosity coefficients are inserted in the Eqs. (7) and (8) and the resulting alignment angle is determined [7]. Thus the system seems to select the alignment angle that minimizes the irreversible energy dissipation rate in accordance with the variational principle. This also means that the energy dissipation rate (8) must be minimal at the preferred alignment angle, θ_0, given by Eq. (7). Thus, the derivative of the function (8) with respect to θ must be zero for θ_0, giving an additional relation between the viscosity coefficients and the alignment angle,

$$\cos 2\theta_0 - = \frac{\tilde{\eta}_2}{2\tilde{\eta}_3} \tag{9}$$

or

$$2\tilde{\eta}_3\tilde{\gamma}_1 + \tilde{\eta}_2\tilde{\gamma}_2 = 0, \tag{10}$$

where θ_0 has been eliminated by using Eq. (7). The Eqs. (7) and (9) do not coincide but they must still give the same value of θ_0. This provides an important cross check for the correctness of the simulation algorithms and experimental methods used to determine the viscosity coefficients and for the computer programs used to run the simulations.

4. Planar elongational flow

4.1 The SLLOD equations of motion for planar elongational flow

A planar irrotational elongational flow arises when an incompressible liquid expands in the x-direction and contracts in the z-direction, see **Figure 4**. Then, the velocity field and the strain rate are given by $\mathbf{u} = \gamma(x\mathbf{e}_x - z\mathbf{e}_z)$ and $\nabla\mathbf{u} = \overline{\nabla\mathbf{u}} = \gamma(\mathbf{e}_x\mathbf{e}_x - \mathbf{e}_z\mathbf{e}_z)$. Planar elongational flow can be studied by applying a special version of the SLLOD equations of motion. Then the problem is that, if the simulation cell is elongated in the x-direction and contracted in the z-direction, the simulation can only continue until the width in the z-direction is equal to twice the range of the interaction potential. However, if the angle between the elongation direction and the x-axis is set to an angle, φ, the periodic lattice of originally quadratic simulation cells is gradually deformed to a lattice of cells shaped like parallelograms. Then, it can be shown that for a special value of this angle, φ_0, the lattice of parallelograms can be remapped onto the original quadratic lattice after a certain time period, so that the simulation becomes continuous, that is, the Kraynik-Reinelt boundary conditions, see **Figure 5** and Refs. [8, 25–28] for details. Then, if the angle between the elongation direction and the x-axis is equal to φ_0, the velocity gradient becomes $\nabla\mathbf{u} = \gamma\left(\mathbf{e}'_x\mathbf{e}'_x - \mathbf{e}'_z\mathbf{e}'_z\right)$, where $\mathbf{e}'_x = \mathbf{e}_x\cos\varphi_0 - \mathbf{e}_z\sin\varphi_0$ and $\mathbf{e}'_z = \mathbf{e}_x\sin\varphi_0 + \mathbf{e}_y\cos\varphi_0$ are the elongation and contraction directions. Inserting this gradient in the SLLOD equations of motion gives,

$$\dot{\mathbf{r}}_i = \frac{\mathbf{p}_i}{m} + \mathbf{r}_i\cdot\nabla\mathbf{u} = \frac{\mathbf{p}_i}{m} + \gamma\mathbf{r}_i\cdot\left(\mathbf{e}'_x\mathbf{e}'_x - \mathbf{e}'_z\mathbf{e}'_z\right) \tag{11a}$$

and

$$\dot{\mathbf{p}}_i = \mathbf{F}_i - \mathbf{p}_i\cdot\nabla\mathbf{u} - \alpha\mathbf{p}_i - \boldsymbol{\beta} = \mathbf{F}_i - \gamma\mathbf{p}_i\cdot\left(\mathbf{e}'_x\mathbf{e}'_x - \mathbf{e}'_z\mathbf{e}'_z\right) - \alpha\mathbf{p}_i - \boldsymbol{\beta}, \tag{11b}$$

where $\mathbf{r}_i$ and $\mathbf{p}_i$ are the position and peculiar momentum, that is, the momentum relative to the macroscopic streaming velocity, of molecule i, $\mathbf{F}_i$ is the force exerted on molecule i by the other molecules, m is the molecular mass, $\mathbf{u}$ is the streaming velocity, γ is the strain rate, $\boldsymbol{\alpha}$ is a thermostting multiplier and $\boldsymbol{\beta}$ is a constraint multiplier used to conserve the linear momentum.

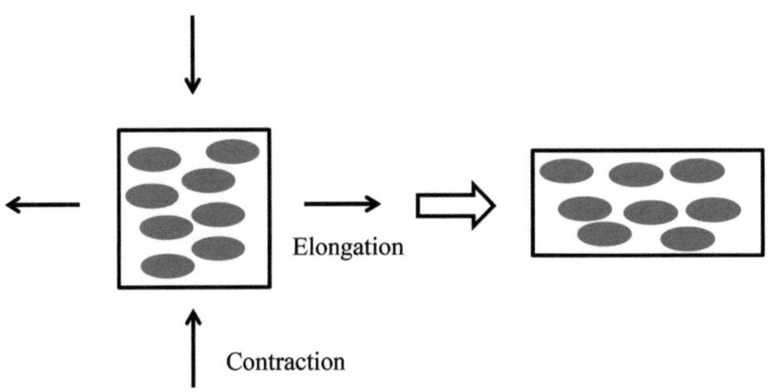

Figure 4.
Schematic representation of a nematic phase of a soft ellipsoid fluid undergoing irrotational extensional flow. The system is elongated in the horizontal direction and contracted in the vertical direction while the volume is constant. The molecules tend to be aligned in the elongation direction.

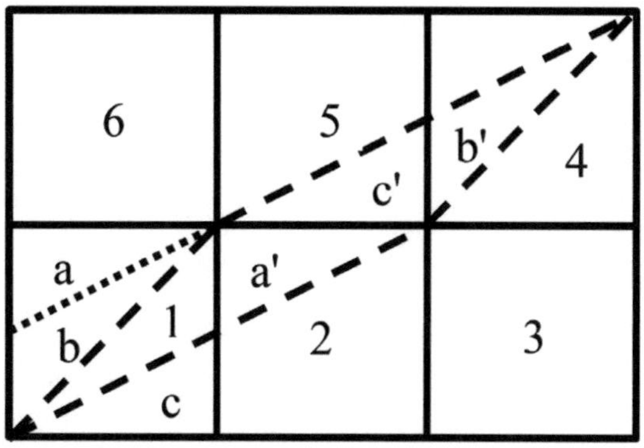

Figure 5.
The Kraynik-Reinelt boundary conditions. The original simulation cell is square 1. When the angle between the elongation direction and the horizontal direction is equal to φ_0, square 1 is deformed to a parallelogram, which, after a given time interval, becomes the dashed parallelogram, partially covering the squares (1–6). Then the triangles a', b', and c' in the parallelogram are periodic copies of the triangles a, b, and c in square 1. If the primed triangles are moved to the corresponding unprimed triangles, a square is recovered and the simulation can continue. Reproduced from Ref. [8], with permission from the PCCP Owner Societies.

4.2 Planar elongational flow of nematic liquid crystals

The director alignment angle is in the first place determined by the mechanical stability in the same way as in shear flow whereby the antisymmetric pressure must be zero. In the linear or Newtonian regime, the alignment angle can be found by using the following relation between the antisymmetric pressure and the strain rate, see Appendix 1,

$$\langle p_2^a \rangle = -\tilde{\gamma}_2 \frac{\gamma}{2} \sin 2\theta, \tag{12}$$

where θ denotes the angle between the director and the elongation direction, and $\tilde{\gamma}_2$ is the cross coupling coefficient between the antisymmetric pressure and the strain rate. From this expression, it is apparent that the alignment angle must be either 0 or 90°, that is, where the torque exerted by the strain rate is equal to zero. For a flow stable calamitic nematic liquid crystal, the cross coupling coefficient $\tilde{\gamma}_2$ is negative [6], so that the 0° orientation parallel to the elongation direction is mechanically stable, and the 90° orientation is unstable.

Just as in planar Couette flow or shear flow, the connection to the variational principle can be made by considering the algebraic expression for the irreversible energy dissipation rate in the linear regime,

$$\dot{w}_{irr} = -\overline{\mathbf{P}} : \overline{\nabla \mathbf{u}} = \left(4\eta + 2\frac{\tilde{\eta}_1}{3} + 2\tilde{\eta}_3 \cos^2 2\theta\right)\gamma^2. \tag{13}$$

If the viscosity coefficient $\tilde{\eta}_3$ is positive, this expression is minimal when θ is equal to 45° but this orientation is excluded because of the mechanical stability (12). If $\tilde{\eta}_3$ on the other hand is negative, this expression attains the same minimal value when θ is equal to 0 or 90°, that is, the elongation or contraction direction. Simulations of a nematic phase of calamitic soft ellipsoids have shown that $\tilde{\eta}_3$ is less than zero [8], so that the energy dissipation rate is minimal in the stable orientation also in this case of planar elongational flow. This is in agreement with the variational principle [1]. See also **Figure 6** where the angular distribution of the director around the elongation direction is displayed.

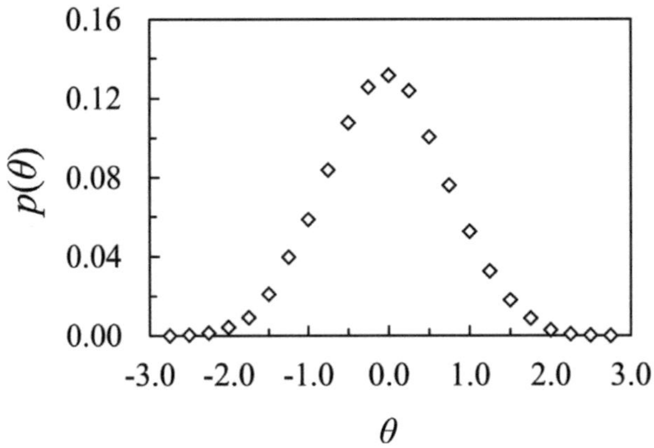

Figure 6.
The angular distribution, $p(\theta)$, of the director of a calamitic nematic liquid crystal consisting of soft ellipsoids around the elongation direction where the angle between the director and the elongation direction is denoted by θ. Reproduced from Ref. [8], with permission from the PCCP Owner Societies.

5. Heat conduction

5.1 Heat flow algorithm

A temperature gradient can be maintained by keeping different regions, 1 and 2, of a system at different temperatures, see **Figure 7**. Mathematically, this can be brought about by adding thermostatting terms for each of the regions 1 and 2 to the ordinary Newtonian equations of motion [29],

$$m\ddot{\mathbf{r}}_i = \mathbf{F}_i - \hat{w}_{1i}\alpha_1 m\dot{\mathbf{r}}_i - \hat{w}_{2i}\alpha_2 m\dot{\mathbf{r}}_i - \zeta, \tag{14}$$

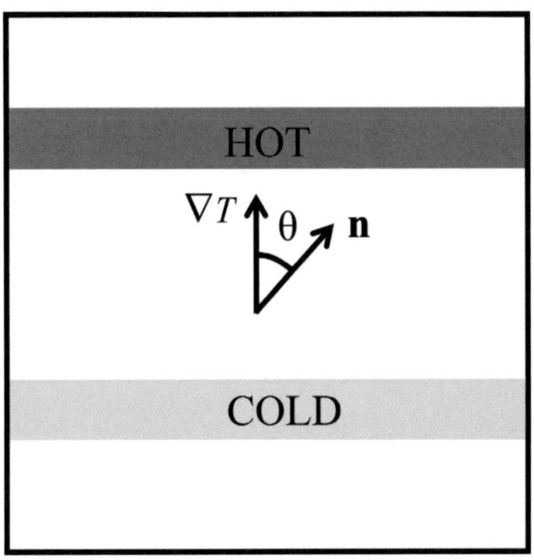

Figure 7.
A temperature gradient is maintained by thermostatting one region (dark gray) of the system at a high temperature and another region (light gray) at a low temperature, whereby heat will flow from the high temperature region to the low temperature region. Reproduced from Ref. [15], with permission from the PCCP Owner Societies.

where m is the molecular mass, $\dot{\mathbf{r}}_i$ and $\ddot{\mathbf{r}}_i$ are the velocity and the acceleration of molecule i, and $\mathbf{F}_i$ is the force exerted on molecule i by the other molecules. The thermostatting terms are $\hat{w}_{1i}\alpha_1 m\dot{\mathbf{r}}_i$ and $\hat{w}_{2i}\alpha_2 m\dot{\mathbf{r}}_i$ where $\hat{w}_{1i}$ and $\hat{w}_{2i}$ are two normalized weight functions. These terms are actually similar to the thermostatting term in Eq. (4b), but here region 1 and region 2 are thermostatted separately at different temperatures. This is achieved by letting the weight functions be Gaussian functions centered in region 1 and 2, respectively, and with decay lengths that are considerably shorter than the distance between these two regions. In this way, only the molecules in region 1 contribute to the temperature in region 1, and only the molecules in region 2 contribute to the temperature in region 2. The molecules far away from the centers of these regions move according to the ordinary Newtonian equations of motion. Note that, it is not necessary to use Gaussian weight functions; it is possible to use any function with a maximum and a rather short decay length. The parameters α_1 and α_2 are thermostatting multipliers in the same way as the multiplier α in Eqs. (4b) and (5), but here, they thermostat the regions 1 and 2 separately. They are determined by applying Gauss's principle of least constraints using the fact that the weighted kinetic energies are constant:

$$\frac{1}{2}\sum_{i=1}^{N}\hat{w}_{1i}m\dot{\mathbf{r}}_i^2 = E_{k1} \tag{15a}$$

and

$$\frac{1}{2}\sum_{i=1}^{N}\hat{w}_{2i}m\dot{\mathbf{r}}_i^2 = E_{k2}, \tag{15b}$$

where E_{k1} and E_{k2} are the weighted kinetic energies for region 1 and 2, respectively. The algebraic expressions for the thermostatting multipliers are given in Ref. [15]. The parameter ζ is a multiplier determined in such a way that the linear momentum of the whole system is constant. It goes to zero in the thermodynamic limit.

5.2 Heat flow in nematic liquid crystals

The heat flow in an axially symmetric system such as a nematic liquid crystal or a cholesteric liquid crystal is given by

$$\langle \mathbf{J}_Q \rangle = -\left[\lambda_{\parallel\ \parallel}\mathbf{nn} + \lambda_{\perp\perp}(\mathbf{1} - \mathbf{nn})\right]\cdot\frac{\nabla T}{T}, \tag{16}$$

where $\langle \mathbf{J}_Q \rangle$ is the heat current density, $\lambda_{\parallel\ \parallel}$ is the heat conductivity parallel to the director of an ordinary achiral nematic liquid crystal or parallel to the cholesteric axis of a cholesteric liquid crystal, $\lambda_{\perp\perp}$ is the heat conductivity perpendicular to the director of a nematic liquid crystal or perpendicular to the cholesteric axis of a cholesteric liquid crystal, T is the absolute temperature, and $\mathbf{n}$ is the director. Then, the irreversible energy dissipation rate of the system due to the heat flow becomes,

$$\begin{aligned}\dot{w}_{irr} = -\langle \mathbf{J}_Q \rangle \cdot \frac{\nabla T}{T} &= \frac{1}{T^2}\left[\lambda_{\perp\perp}\nabla T\cdot\nabla T + \left(\lambda_{\parallel\ \parallel} - \lambda_{\perp\perp}\right)(\mathbf{n}\cdot\nabla T)^2\right] \\ &= \left|\tfrac{\partial_z T}{T}\right|^2\left[\lambda_{\perp\perp} + \left(\lambda_{\parallel\ \parallel} - \lambda_{\perp\perp}\right)\cos^2\theta\right],\end{aligned} \tag{17}$$

where the last equality has been obtained by assuming that the director lies in the zx-plane forming an angle θ with the temperature gradient, see **Figure 8**. When

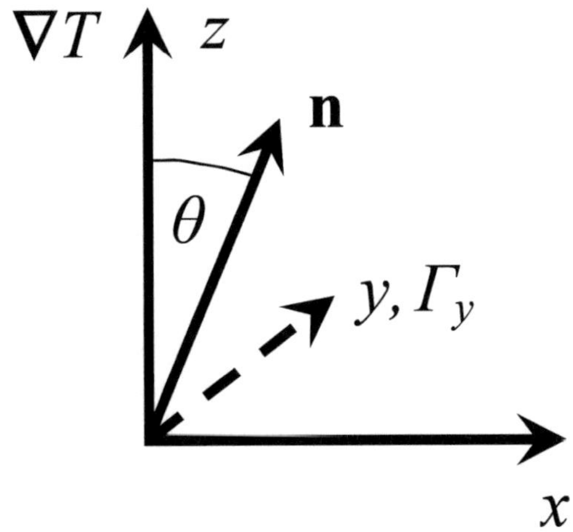

Figure 8.
*A schematic view of a nematic liquid crystal subject to a temperature gradient is shown. The temperature gradient T points in the z-direction, and the director **n** lies in the zx-plane forming an angle θ to the z-axis. Then a torque **Γ** arises in the direction of the y-axis. Reproduced from Ref. [15], with permission from the PCCP Owner Societies.*

$\lambda_{\parallel\,\parallel} > \lambda_{\perp\perp}$, as in a nematic liquid crystal consisting of calamitic molecules, the heat current density and thereby $\dot{w}_{irr}$ are maximal when the temperature gradient and the director are parallel and minimal when they are perpendicular to each other. Conversely, when $\lambda_{\parallel\,\parallel} < \lambda_{\perp\perp}$, as in a nematic liquid crystal consisting of discotic molecules, the heat current density and the irreversible energy dissipation rate are maximal when the director is perpendicular to the temperature gradient and minimal when it is parallel to the temperature gradient.

The temperature gradient exerts a torque on the molecules around an axis perpendicular to itself and perpendicular to the director, see **Figure 8**. This torque must be zero in the parallel and perpendicular orientations due to the symmetry, but it is impossible to determine whether these orientations are stable or unstable. Unfortunately, there is no linear relation between the torque and the temperature gradient since they are pseudovectors and polar vectors, respectively, due to the axial symmetry of the system. However, a quantitative relation between them can be obtained by noting that a cross coupling between a pseudo vector and a symmetric second rank tensor is allowed. The latter quantity can be obtained by forming the dyadic product of the temperature gradient, giving the following relation [15],

$$\mathbf{\Gamma} = \mu \boldsymbol{\varepsilon} : \mathbf{nn} \cdot \frac{\nabla T}{T}\frac{\nabla T}{T} = \mu \mathbf{n} \cdot \frac{\nabla T}{T} \mathbf{n} \times \frac{\nabla T}{T} = \mu \left| \frac{\partial_z T}{T} \right|^2 \cos\theta \sin\theta \, \mathbf{e}_y = \frac{1}{2}\mu \left| \frac{\partial_z T}{T} \right|^2 \sin 2\theta \, \mathbf{e}_y, \tag{18}$$

where $\mathbf{\Gamma}$ is the torque density, μ is a cross coupling coefficient, and $\boldsymbol{\varepsilon}$ is the Levi-Civita tensor. The third equality is obtained by assuming that the temperature gradient points in the z-direction, and the director lies in the zx-plane, see **Figure 8**, whereby θ becomes the angle between these two vectors. This relation fulfills the symmetry condition according to which the torque must be zero when the director is parallel or perpendicular to the temperature gradient. Moreover, the torque is proportional to the square of the temperature gradient for given angle θ. Note also that, this relation is analogous to the relation between the strain rate and the antisymmetric pressure in planar elongational flow (12).

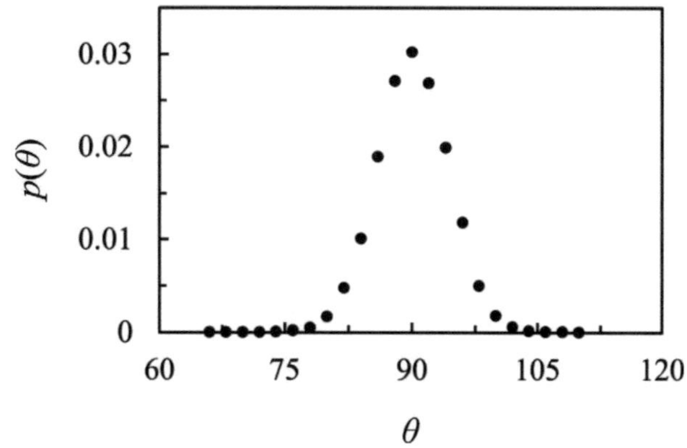

Figure 9.
The angular distribution, $p(\theta)$, of the director of a nematic liquid crystal consisting of soft calamitic ellipsoids around the temperature gradient where the angle between the director and the temperature gradient is denoted by θ. Reproduced from Ref. [15], with permission from the PCCP Owner Societies.

The director orientation can be determined by simulating systems, where a temperature gradient and a heat flow are maintained by thermostatting different parts of the system at different temperatures by using the above simulation algorithm (14). Such simulations have shown that the director of nematic liquid crystals consisting of soft calamitic ellipsoids tends to align perpendicularly to the temperature gradient, see **Figure 9**, whereas the director of nematic liquid crystals consisting of discotic ellipsoids tends to align parallel to the temperature gradient. Thus, the energy dissipation rate is minimal in both cases. Moreover, if the director is constrained to attain a fixed orientation between the perpendicular and parallel orientation relative to temperature gradient by applying the Lagrangian constraint algorithm (3), the torque exerted can be obtained. Then, it is found that this torque turns the director of a calamitic system toward the perpendicular orientation and the director of a discotic system toward the parallel orientation. The same orientation behavior of the directors of calamitic and discotic nematic liquid crystals relative to the temperature gradient was observed in an earlier work [21]. However, then the Evans heat flow algorithm [22] was applied where a fictitious external field under non-Newtonian equations of motion rather than a real temperature gradient drives the heat flow. Therefore, it was not possible to determine whether the orientation phenomena were a real effect or a consequence of the non-Newtonian synthetic equations of motion.

There are also some early experimental works on the orientation of the director of nematic liquid crystals relative to temperature gradients [9–14] that probably support the conclusions of these heat flow simulations. Unfortunately, it is very difficult to carry out these experiments because if the temperature gradient is too large, there will be convection in the system, and if the temperature gradient is too small, it will not be strong enough to overcome the elastic torques or the surface torques. Therefore, these experiments are not fully conclusive.

Finally, one example where the director orientation relative to a temperature gradient definitely is the one that minimizes the irreversible energy dissipation rate is a cholesteric liquid crystal. In this system, the director rotates in space around the cholesteric axis forming a spiral structure. Then experimental studies, where a temperature gradient is applied, have shown that the cholesteric axis orients parallel to the temperature gradient, whereby the energy dissipation rate is minimized since the heat conductivity is greater in the direction perpendicular to the cholesteric axis than in the parallel direction. Moreover, the whole spiral structure starts rotating in time. This phenomenon is known as thermomechanical coupling [3, 4, 18–20, 30, 31].

There are quite a few experimental studies available on this phenomenon, where it has been found in a conclusive way that the cholesteric axis remains parallel to the temperature gradient, so this orientation seems to be stable, and thus the irreversible energy dissipation rate is minimal.

We can consequently conclude that the orientation of the director relative to the temperature gradient is consistent with the variational principle [1] even though the coupling between the torque and the temperature gradient is quadratic rather than linear and the system is inhomogeneous. However, the temperature gradient is rather weak, so we still remain in the linear regime.

6. Effects of the thermostat

In the above simulations of shear flow and elongational flow, the velocity gradient does work on the system that is converted to heat, which must be removed in order to keep the temperature constant and to maintain a steady state. In a real macroscopic system, this takes place by heat conduction to the system boundaries and this could in principle be arranged in a microscopic simulation cell as well. Unfortunately, this is inconvenient because a temperature gradient of molecular dimensions would make the system inhomogeneous, and thus make it difficult to define the thermodynamic state. Therefore, the temperature is kept constant by forcing the kinetic energy to be a constant of motion by applying a Gaussian thermostat, see Eq. (5). This thermostat was originally devised independently by Hoover *et al.* [32–34] and by Evans [22]. The equilibrium ensemble averages of the phase functions and time correlation functions generated when this thermostat is applied are essentially canonical [35]. Away from equilibrium, it can be shown that the effect of the Gaussian thermostat on the ensemble averages is proportional to the square of the external field, whereas the thermodynamic fluxes driven by the field are directly proportional to the field in a linear transport process. Thus, the corresponding linear transport coefficients that are equal to the ratio of the flux and the field in the limit of zero field are independent of the thermostat. Therefore, transport coefficients obtained from the simulations of shear flow and elongational flow are independent of the thermostat since there is a linear relation between the velocity gradient and the shear stress in the Newtonian regime and since we are interested in the limit of zero velocity gradient. Neither is the correctness of the variational principle affected by the thermostat since it is valid in the linear regime.

The situation is different in the heat flow simulations because here we actually want a temperature gradient. This gradient is obtained by applying two bar thermostats at different temperatures acting over a limited range and separated by a distance that is long compared to this range, see **Figure** 7 and Eq. (14). Therefore, the movements of only a small fraction of the molecules are affected by the thermostats, whereas the movements of the majority of the molecules away from the bar thermostats are governed by the ordinary Newtonian equations of motion. Thus, it is reasonable to assume that the influence of the details of the thermostat on the ensemble averages of the phase functions is limited in this case too.

7. Conclusion

The purpose of this work has been to test a recently proven variational principle according to which the irreversible energy dissipation rate is minimal in the linear regime of a nonequilibrium steady state. Therefore, we have reviewed molecular dynamics simulations and experimental work on director orientation phenomena in nematic liquid crystals and in cholesteric liquid crystals under external dissipative fields such as velocity gradients and temperature gradients. A general observation that we have made is that in all the examples studied, the director of the liquid crystals seems to attain precisely that alignment angle relative to the external dissipative field that minimizes the irreversible energy dissipation rate.

In a nematic liquid crystal, the director orientation is in the first place determined by a mechanical stability criterion, namely, that the external torques acting on the system must be zero at mechanical equilibrium. This makes it possible to derive an exact relation between the alignment angle relative to the streamlines and the viscosity coefficients in the linear or Newtonian regime of planar elongational flow and of planar Couette flow. Both simulations and experimental measurements imply that the irreversible energy dissipation rate is minimal at this mechanically stable orientation.

It can be shown that the elongation direction is the stable orientation of flow stable calamitic nematic liquid crystals undergoing elongational flow in the linear regime. It can also be shown that the value of the energy dissipation rate is the same in the contraction direction and in the elongation direction, and that this value is either the maximal or the minimal value by using the linear phenomenological relations between the strain rate and the pressure. Simulations of the calamitic soft ellipsoid fluid have shown that the irreversible energy dissipation rate is minimal in the elongation direction.

In calamitic nematic liquid crystals, the heat conductivity is larger in the direction parallel to the director than in the perpendicular direction, and the reverse is true for discotic nematic liquid crystals. Thus, the irreversible energy dissipation rate due to the heat flow depends on the angle between the director and the temperature gradient. When a nematic liquid crystal is subjected to a temperature gradient, a torque is exerted on the molecules. Due to symmetry, this torque must be proportional to the square of the temperature gradient and it must be zero when the director is parallel or perpendicular to this gradient.

In simulations of nematic phases of soft ellipsoids under a temperature gradient, it turns out that the director of a calamitic nematic liquid crystal aligns perpendicularly to the temperature gradient, whereas the director of a discotic nematic liquid crystal attains the parallel orientation. In both cases, the irreversible energy dissipation rate is minimal. These simulation results are probably supported by some experimental measurements, but they are difficult to carry out in practice so they are not fully conclusive.

Finally, one system where there is definitely a conclusive experimental evidence for the fact that the director attains the orientation that minimizes the energy dissipation rate due to a temperature gradient is the cholesteric liquid crystal. The cholesteric axis of droplets of cholesteric liquid crystals orient parallel to a temperature gradient and the director rotates. This is a well-established phenomenon observed in studies of thermomechanical coupling, and since the heat conductivity is lower in the direction of the cholesteric axis than in the perpendicular direction, the energy dissipation rate is minimal in this case.

Thus, the director orientation relative to a temperature gradient also follows the variational principle even though there is a quadratic coupling between the torque and the temperature gradient. However, the temperature gradients are rather low so we are still in the linear regime.

Acknowledgements

We gratefully acknowledge financial support from the Knut and Alice Wallenberg Foundation (Project number KAW 2012.0078) and Vetenskapsrådet (Swedish Research Council) (Project number 2013-5171). The simulations were performed using resources provided by the Swedish National Infrastructure for Computing (SNIC) at PDC, HPC2N, and NSC.) We also acknowledge PRACE for awarding us access to Hazelhen based in Germany at Rechenzentrum Stuttgart.

A. Appendix 1: Relation between the pressure tensor, velocity gradient, and viscosity coefficients

The relation between the velocity gradient, $\nabla\mathbf{u}$, and the pressure tensor, $\mathbf{P}$, is more complicated in an axially symmetric system such as nematic liquid crystal than in an isotropic fluid due to the lower symmetry. In order to derive the linear phenomenological relations between the velocity gradient and the pressure, it is appropriate to begin by identifying the thermodynamic forces and fluxes in the expression for the irreversible entropy production [3, 4, 23, 36]:

$$\sigma = -\frac{1}{T}\left\{\overline{\mathbf{P}} : \overline{\nabla\mathbf{u}} + 2\mathbf{P}^{a}\cdot(½\nabla\times\mathbf{u}-\boldsymbol{\Omega}) + \left(\frac{1}{3}\mathrm{Tr}(\mathbf{P}) - p_{eq}\right)\nabla\cdot\mathbf{u}\right\}, \tag{A.1}$$

where T is the absolute temperature, and $\mathbf{u}$ is the streaming velocity. The various parts of the second rank tensor are denoted in the following manner: the symmetric traceless part is given by $\overline{\mathbf{A}} = ½\left(\mathbf{A}+\mathbf{A}^{T}\right) - (1/3)\mathrm{Tr}(\mathbf{A})\mathbf{1}$ and the pseudovector dual of the antisymmetric part is denoted by $\mathbf{A}^{a} = -½\boldsymbol{\varepsilon}{:}\mathbf{A} = -½\varepsilon_{\alpha\beta\gamma}A_{\gamma\beta}$, where $\boldsymbol{\varepsilon}$ is the Levi-Civita tensor. Three pairs of thermodynamic forces and fluxes can be identified by inspection of the irreversible entropy production, namely, the symmetric traceless pressure tensor and the traceless strain rate, $\overline{\mathbf{P}}$ and $\overline{\nabla\mathbf{u}}$, the antisymmetric pressure and the difference between the rotation and the director angular velocity, $\mathbf{P}^{a}$ and $½\nabla\times\mathbf{u}-\boldsymbol{\Omega}$, and the difference between the trace of the pressure tensor and the equilibrium pressure of a quiescent liquid crystal, and the trace of the strain rate, $(1/3)\mathrm{Tr}(\mathbf{P}) - p_{eq}$ and $\nabla\cdot\mathbf{u}$. Note that the strain rate is defined as $½\left[\nabla\mathbf{u}+(\nabla\mathbf{u})^{T}\right]$, and it is always symmetric. In a uniaxially symmetric nematic liquid crystal, the relations between the pressure and the velocity gradient can be deduced by symmetry arguments, and they can be expressed in a few different equivalent ways [23, 36]. It has been found that a notation due to Hess [36] is the most convenient one for deducing Green-Kubo relations and NEMD-algorithms:

$$\langle\overline{\mathbf{P}}\rangle = -2\eta\overline{\nabla\mathbf{u}} - \tilde{\eta}_1\overline{\overline{\mathbf{nn}}\cdot\overline{\nabla\mathbf{u}}} - 2\tilde{\eta}_3\overline{\mathbf{nn}}\,\overline{\mathbf{nn}}{:}\overline{\nabla\mathbf{u}} + 2\tilde{\eta}_2\overline{\overline{\mathbf{nn}}\cdot\boldsymbol{\varepsilon}\cdot(½\nabla\times\mathbf{u}-\boldsymbol{\Omega})} - \zeta\overline{\mathbf{nn}}\nabla\cdot\mathbf{u}, \tag{A.2a}$$

$$\langle\mathbf{P}^{a}\rangle = -\frac{\tilde{\gamma}_1}{2}(½\nabla\times\mathbf{u}-\boldsymbol{\Omega}) - \frac{\tilde{\gamma}_2}{2}\boldsymbol{\varepsilon} : \left(\overline{\mathbf{nn}}\cdot\overline{\nabla\mathbf{u}}\right), \tag{A.2b}$$

and

$$\frac{1}{3}\langle\mathrm{Tr}(\mathbf{P})\rangle - p_{eq} = -\eta_V\nabla\cdot\mathbf{u} - \kappa\overline{\mathbf{nn}}{:}\overline{\nabla\mathbf{u}}, \tag{A.2c}$$

where the products involving the Levi-Civita tensor $\boldsymbol{\varepsilon}$ are defined in the following way: $\boldsymbol{\varepsilon}:\mathbf{A}=\varepsilon_{\alpha\beta\gamma}A_{\gamma\beta}$ and $\mathbf{A}\cdot\boldsymbol{\varepsilon}\cdot\mathbf{B}=A_{\alpha\beta}\varepsilon_{\beta\gamma\delta}B_{\delta}$. The quantities η, $\tilde{\eta}_1$, and $\tilde{\eta}_3$ are shear viscosities, $\tilde{\gamma}_1$ is the twist viscosity, η_V is the volume viscosity, $\tilde{\eta}_2$ is the cross coupling coefficient relating the difference between the rotation and the director angular velocity, and the symmetric traceless pressure. According to the Onsager reciprocity relations (ORR), this coefficient is equal to $\tilde{\gamma}_2/2$, the cross coupling coefficient relating the traceless strain rate and the antisymmetric pressure. The trace of the strain rate and the symmetric traceless pressure are related by the cross coupling coefficient, ζ, which, according to the ORR, is equal to the cross coupling coefficient κ between the traceless strain rate and the difference between the trace of the pressure tensor and the equilibrium pressure.

Application of a planar Couette velocity gradient, $\nabla\mathbf{u}=\gamma\,\mathbf{e}_z\mathbf{e}_x$, where $\gamma=\partial_z u_x$ is the shear rate and fixation of the director in the zx-plane at an angle θ relative to the stream lines, see **Figure 2**, by application of an electric or a magnetic field gives the following relations between the pressure tensor components and the strain rate in a director based coordinate system $(\mathbf{e}_1,\mathbf{e}_2,\mathbf{e}_3)$ where the director points in the $\mathbf{e}_3$-direction:

$$\langle\overline{p}_{11}\rangle=\left(\eta+\frac{\tilde{\eta}_3}{3}\right)\gamma\sin 2\theta, \tag{A.3a}$$

$$\langle\overline{p}_{22}\rangle=\frac{1}{3}(\tilde{\eta}_1+\tilde{\eta}_3)\gamma\sin 2\theta, \tag{A.3b}$$

$$\langle\overline{p}_{33}\rangle=-\left(\eta+\frac{\tilde{\eta}_1}{3}+2\frac{\tilde{\eta}_3}{3}\right)\gamma\sin 2\theta, \tag{A.3c}$$

$$\langle\overline{p}_{31}\rangle=\left(\eta+\frac{\tilde{\eta}_1}{6}\right)\gamma\cos 2\theta+\tilde{\eta}_2\frac{\gamma}{2}, \tag{A.3d}$$

and

$$2\langle p_2^a\rangle=\langle\hat{\lambda}_2\rangle=-\tilde{\gamma}_1\frac{\gamma}{2}-\tilde{\gamma}_2\frac{\gamma}{2}\cos 2\theta, \tag{A.3e}$$

where $\hat{\lambda}_2$ is the external torque density acting on the system. From these equations, it is apparent that the various elements of the pressure tensor are linear functions of $\sin 2\theta$ and $\cos 2\theta$, so the various viscosity coefficients can be evaluated by fixing the director at a few different angles relative to the stream lines and calculating the averages of the pressure tensor elements.

In a planar elongational flow [8, 26–28], where the elongation direction is parallel to the x-axis, the contraction direction is parallel to the negative z-axis, and the velocity field is equal to $\mathbf{u}=\gamma(x\mathbf{e}_x-z\mathbf{e}_z)$, so that the velocity gradient becomes $\nabla\mathbf{u}=\gamma(\mathbf{e}_x\mathbf{e}_x-\mathbf{e}_z\mathbf{e}_z)$. Then the linear relations between the velocity gradient and the pressure become the following in a director-based coordinate system $(\mathbf{e}_1,\mathbf{e}_2,\mathbf{e}_3)$ where the director points in the $\mathbf{e}_1$-direction, and θ is the angle between the director and the elongation direction or x-axis, $\mathbf{e}_2=\mathbf{e}_y$ and $\mathbf{e}_3=\mathbf{e}_1\times\mathbf{e}_2$,

$$\langle\overline{p}_{11}\rangle=-2\left(\eta+\frac{\tilde{\eta}_3}{3}\right)\gamma\cos 2\theta, \tag{A.4a}$$

$$\langle\overline{p}_{22}\rangle=-\frac{2}{3}(\tilde{\eta}_1+\tilde{\eta}_3)\gamma\cos 2\theta, \tag{A.4b}$$

$$\langle \overline{p}_{33} \rangle = 2\left(\eta + \frac{\tilde{\eta}_1}{3} + 2\frac{\tilde{\eta}_3}{3}\right)\gamma \cos 2\theta, \tag{A.4c}$$

$$\langle \overline{p}_{31} \rangle = \left(2\eta + \frac{\tilde{\eta}_1}{3}\right)\gamma \sin 2\theta, \tag{A.4d}$$

and

$$2\langle p_2^a \rangle = \langle \hat{\lambda}_2 \rangle = -\tilde{\gamma}_2 \gamma \sin 2\theta. \tag{A.4e}$$

If these expressions for the pressure tensor are inserted in the expression for energy dissipation rate (A.1), we obtain

$$\begin{aligned} \dot{w}_{irr} &= \overline{\mathbf{P}} : \overline{\nabla \mathbf{u}} = -\left\langle \tfrac{1}{2}\left(\overline{p}_{33}^s - \overline{p}_{11}^s\right) \sin 2\theta - \overline{p}_{31}^s \cos 2\theta \right\rangle_\gamma \\ &= \left(\eta + \frac{\tilde{\eta}_1}{6} + \frac{\tilde{\eta}_3}{2} \sin^2 2\theta + \frac{\tilde{\eta}_2}{2} \cos 2\theta\right)\gamma^2, \end{aligned} \tag{A.5}$$

for planar Couette flow and

$$\dot{w}_{irr} = \left(4\eta + 2\frac{\tilde{\eta}_1}{3} + 2\tilde{\eta}_3 \sin^2 2\theta\right)\gamma^2 \tag{A.6}$$

for planar elongational flow. The subscript γ denotes that the average is evaluated in a nonequilibrium ensemble at a finite shear rate.

B. Appendix 2: The Gay-Berne potential

In order to evaluate the above expressions for the irreversible work in shear flow, elongational flow, and heat flow, we have simulated a coarse grained model system composed of molecules interacting via a purely repulsive version of the commonly used Gay-Berne potential [16, 17, 21],

$$U(\mathbf{r}_{12}, \hat{\mathbf{u}}_1, \hat{\mathbf{u}}_2) = 4\varepsilon(\hat{\mathbf{r}}_{12}, \hat{\mathbf{u}}_1, \hat{\mathbf{u}}_2)\left(\frac{\sigma_0}{r_{12} - \sigma(\hat{\mathbf{r}}_{12}, \hat{\mathbf{u}}_1, \hat{\mathbf{u}}_2) + \sigma_0}\right)^{18}, \tag{A.7}$$

where $\mathbf{r}_{12} = \mathbf{r}_2 - \mathbf{r}_1$ is the distance vector from the center of mass of molecule 1 to the center of mass of molecule 2, $\hat{\mathbf{r}}_{12}$ is the unit vector in the direction of $\mathbf{r}_{12}$, r_{12} is the length of the vector $\mathbf{r}_{12}$, and $\hat{\mathbf{u}}_1$ and $\hat{\mathbf{u}}_2$ are the unit vectors parallel to the axes of revolution of molecule 1 and molecule 2. The parameter σ_0 is the length of the axis perpendicular to the axis of revolution, that is, the minor axis of a calamitic ellipsoid of revolution and the major axis of a discotic ellipsoid of revolution. The strength and range parameters are given by

$$\begin{aligned} \varepsilon(\hat{\mathbf{r}}_{12}, \hat{\mathbf{u}}_1, \hat{\mathbf{u}}_2) = \varepsilon_0 &\left[1 - \chi^2(\hat{\mathbf{u}}_1 \cdot \hat{\mathbf{u}}_2)^2\right]^{-1/2} \\ &\left\{1 - \frac{\chi'}{2}\left[\frac{(\hat{\mathbf{r}}_{12} \cdot \hat{\mathbf{u}}_1 + \hat{\mathbf{r}}_{12} \cdot \hat{\mathbf{u}}_2)^2}{1 + \chi' \hat{\mathbf{u}}_1 \cdot \hat{\mathbf{u}}_2} + \frac{(\hat{\mathbf{r}}_{12} \cdot \hat{\mathbf{u}}_1 - \hat{\mathbf{r}}_{12} \cdot \hat{\mathbf{u}}_2)^2}{1 - \chi' \hat{\mathbf{u}}_1 \cdot \hat{\mathbf{u}}_2}\right]\right\}^2 \end{aligned} \tag{A.8a}$$

and

$$\sigma(\hat{\mathbf{r}}_{12}, \hat{\mathbf{u}}_1, \hat{\mathbf{u}}_2) = \sigma_0 \left\{ 1 - \frac{\chi}{2} \left[\frac{(\hat{\mathbf{r}}_{12} \cdot \hat{\mathbf{u}}_1 + \hat{\mathbf{r}}_{12} \cdot \hat{\mathbf{u}}_2)^2}{1 + \chi \hat{\mathbf{u}}_1 \cdot \hat{\mathbf{u}}_2} + \frac{(\hat{\mathbf{r}}_{12} \cdot \hat{\mathbf{u}}_1 - \hat{\mathbf{r}}_{12} \cdot \hat{\mathbf{u}}_2)^2}{1 - \chi \hat{\mathbf{u}}_1 \cdot \hat{\mathbf{u}}_2} \right] \right\}^{-1/2}, \tag{A.8b}$$

where the parameter χ is equal to $(\kappa^2 - 1)/(\kappa^2 + 1)$ and κ is the ratio of the axis of revolution and the axis perpendicular to this axis, χ' is equal to $\left(\kappa'^{1/2} - 1\right)/\left(\kappa'^{1/2} + 1\right)$ and κ' is the ratio of the potential energy minima of the side by side and end to end configurations of calamitic ellipsoids or the ratio of the edge-to-edge and face-to-face configurations of discotic ellipsoids, and ε_0 denotes the depth of the potential minimum in the cross configuration, where $\hat{\mathbf{r}}_{12}$, $\hat{\mathbf{u}}_1$, and $\hat{\mathbf{u}}_2$ are perpendicular to each other. The parameters κ and κ' have been given the values 3 and 5, respectively, for the calamitic ellipsoids and 1/3 and 1/5 for the discotic ellipsoids.

The denominators in Eqs. (A.8a) and (A.8b) are never equal to zero because the absolute value of the scalar product $\hat{\mathbf{u}}_1 \cdot \hat{\mathbf{u}}_2$ is less than or equal to one since $\hat{\mathbf{u}}_1$ and $\hat{\mathbf{u}}_2$ are unit vectors, and the absolute values of the parameters χ and χ' are less than one. The ordinary Lennard-Jones potential is recovered in the limit when κ and κ' go to one. Note that, the potential is purely repulsive, so there are no potential minima but the value of κ' optimized for the attractive Gay-Berne potential has been retained. The transport properties of this system of purely repulsive soft ellipsoids are similar to those of a system where the molecules interact according to the conventional Gay-Berne potential with attraction as well, so the results are still relevant.

Author details

Sten Sarman*, Yonglei Wang and Aatto Laaksonen
Department of Materials and Environmental Chemistry, Arrhenius Laboratory, Stockholm University, Stockholm, Sweden

*Address all correspondence to: sarman@ownit.nu

References

[1] Evans DJ, Searles DJ, Williams SR. Fundamentals of Classical Statistical Thermodynamics: Dissipation, Relaxation and Fluctuation Theorems. Berlin: Wiley-VCH; 2016

[2] de Groot SR, Mazur P. Nonequilibrium Thermodynamics. New York: Dover; 1984

[3] Chandrasekhar S. Liquid Crystals. Cambridge: Cambridge University Press; 1992

[4] de Gennes PG, Prost J. The Physics of Liquid Crystals. Oxford: Clarendon Press; 1993

[5] Sarman S. Microscopic theory of liquid crystal rheology. The Journal of Chemical Physics. 1995;**103**:393

[6] Sarman S. Nonequilibrium molecular dynamics of liquid crystal shear flow. The Journal of Chemical Physics. 1995; **103**:10378

[7] Jadzyn J, Czechowski G. The shear viscosity minimum of freely flowing nematic liquid crystals. Journal of Physics: Condensed Matter. 2001;**13**:L261

[8] Sarman S, Laaksonen A. Molecular dynamics simulation of planar elongational flow in a nematic liquid crystal based on the Gay–Berne potential. Physical Chemistry Chemical Physics. 2015;**17**:3332

[9] Stewart GW. X-Ray diffraction intensity of the two liquid phases of para-azoxyanisol. The Journal of Chemical Physics. 1936;**4**:231

[10] Stewart GW, Holland DO, Reynolds LM. Orientation of liquid crystals by heat conduction. Physics Review. 1940; **58**:174

[11] Picot JJC, Fredrickson AG. Interfacial and electrical effects on thermal conductivity of nematic liquid crystals. Industrial and Engineering Chemistry Fundamentals. 1968;**7**:84

[12] Fisher J, Fredrickson AG. Transport processes in anisotropic fluids II. Coupling of momentum and energy transport in a nematic mesophase. Molecular Crystals and Liquid Crystals. 1969;**6**:255

[13] Patharkar MN, Rajan VSV, Picot JJC. Interfacial and temperature gradient effects on thermal conductivity of a liquid crystal. Molecular Crystals and Liquid Crystals. 1971;**15**:225

[14] Currie PK. The orientation of liquid crystals by temperature gradients. Rheologica Acta. 1973;**12**:165

[15] Sarman S, Laaksonen A. Director alignment relative to the temperature gradient in nematic liquid crystals studied by molecular dynamics simulation. Physical Chemistry Chemical Physics. 2014;**16**:14741

[16] Gay JG, Berne BJ. Modification of the overlap potential to mimic a linear site–site potential. The Journal of Chemical Physics. 1981;**74**:3316

[17] Bates MA, Luckhurst GR. Computer simulation studies of anisotropic systems. XXVI. Monte Carlo investigations of a Gay–Berne discotic at constant pressure. The Journal of Chemical Physics. 1996;**104**:6696

[18] Éber N, Jánossy I. An experiment on the thermomechanical coupling in cholesterics. Molecular Crystals and Liquid Crystals Science and Technology. Section A. Molecular Crystals and Liquid Crystals. 1982;**72**:233

[19] Oswald P, Dequidt A. Measurement of the continuous Lehmann rotation of cholesteric droplets subjected to a

temperature gradient. Physical Review Letters. 2008;**100**:217802

[20] Oswald P. Microscopic vs. macroscopic origin of the Lehmann effect in cholesteric liquid crystals. European Physical Journal E: Soft Matter and Biological Physics. 2012;**35**:10

[21] Sarman S. Molecular dynamics of heat flow in nematic liquid crystals. The Journal of Chemical Physics. 1994;**101**:480

[22] Evans DJ, Morriss GP. Statistical Mechanics of Nonequilibrium Liquids. London: Academic Press; 1990

[23] Leslie FM. Some constitutive equations for anisotropic fluids. The Quarterly Journal of Mechanics and Applied Mathematics. 1966;**19**:357

[24] Sarman S, Laaksonen A. Flow alignment phenomena in liquid crystals studied by molecular dynamics simulation. The Journal of Chemical Physics. 2009;**131**:144904

[25] Kraynik AM, Reinelt DA. Extensional motions of spatially periodic lattices. International Journal of Multiphase Flow. 1992;**18**:1045

[26] Baranyai A, Cummings PT. Nonequilibrium molecular dynamics study of shear and shear-free flows in simple fluids. The Journal of Chemical Physics. 1995;**103**:10217

[27] Todd BD, Daivis PJ. Nonequilibrium molecular dynamics simulations of planar elongational flow with spatially and temporally periodic boundary conditions. Physical Review Letters. 1998;**81**:1118

[28] Todd BD, Daivis PJ. Homogeneous non-equilibrium molecular dynamics simulations of viscous flow: Techniques and applications. Molecular Simulation. 2007;**33**:189

[29] Ikeshoji T, Hafskjold B. Non-equilibrium molecular dynamics calculation of heat conduction in liquid and through liquid-gas interface. Molecular Physics. 1994;**81**:251

[30] Leslie FM. Some thermal effects in cholesteric liquid crystals. Proceedings of the Royal Society of London. Series A, Mathematical and Physical Sciences. 1968;**307**:359

[31] Leslie FM. Thermo-mechanical coupling in cholesteric liquid crystals. Symposium of the Faraday Society. 1971;**5**:33

[32] Hoover WG, Ladd AJC, Moran B. High-strain-rate plastic flow studied via nonequilibrium molecular dynamics. Physical Review Letters. 1982;**48**:1818

[33] Evans DJ, Hoover WG, Failor BH, Moran B, Ladd AJC. Nonequilibrium molecular dynamics via Gauss's principle of least constraint. Physical Review A. 1983;**28**:1016

[34] Hoover WG. Computational Statistical Mechanics. Burlington, MA: Elsevier; 1991

[35] Evans DJ, Sarman S. Equivalence of thermostatted nonlinear responses. Physical Review E. 1993;**48**:65

[36] Hess S. Transport phenomena in anisotropie fluids and liquid crystals. Journal of Non-Equilibrium Thermodynamics. 1986;**11**:175

Chapter 6

Equilibrium and Nonequilibrium Hydrodynamic Modes of a Nematic Liquid Crystal

Jorge Fernando Camacho and Rosalío Fernando Rodríguez

Abstract

We use a fluctuating hydrodynamics (*FH*) approach to study the fluctuations of the hydrodynamic variables of a thermotropic nematic liquid crystal (*NLC*) in a nonequilibrium steady state (*NESS*). This *NESS* is produced by an externally imposed temperature gradient and a uniform gravity field. We calculate analytically the equilibrium and nonequilibrium seven modes of the *NLC* in this *NESS*. These modes consist of a pair of sound modes, one orientation mode of the director and two visco-heat modes formed by the coupling of the shear and thermal modes. We find that the nonequilibrium effects produced by the external gradients only affect the longitudinal modes. The analytic expressions for the visco-heat modes show explicitly how the heat and shear modes of the *NLC* are coupled. We show that they may become propagative, a feature that also occurs in the simple fluid and suggests the realization of new experiments. We show that in equilibrium and in the isotropic limit of the *NLC*, our modes reduce to well-known results in the literature. For the *NESS* considered, we point out the differences between our modes and those reported by other authors. We close the chapter by proposing the calculation of other physical quantities that lend themselves to a more direct comparison with possible experiments for this system.

Keywords: fluctuating hydrodynamics, nonequilibrium fluctuations, hydrodynamic modes, thermotropic nematic liquid crystals

1. Introduction

When a fluid is in thermodynamic equilibrium, its state variables always present spontaneous microscopic fluctuations due to the thermal excitations of its molecules, producing deviations around the state of equilibrium. The theory of fluctuations for fluids in states close to equilibrium was initiated long ago by Einstein and Onsager, and it has been reformulated in several but equivalent ways. The first more systematic approach to introduce thermal fluctuations into the hydrodynamic equations was the fluctuating hydrodynamics (*FH*) of Landau and Lifshitz [1, 2]. It stems from the idea that the hydrodynamic equations are valid for any flow, including fluctuating changes in its state. Accordingly, stochastic currents are incorporated into the deterministic energy and momentum fluxes by adding fluctuating sources. This theory was put on a firm basis within the framework of the general theory of stationary Gaussian Markov processes by Fox and Uhlenbeck [3–5].

This approach has matched the theory of Onsager and Machlup with that of Landau and Lifshitz for systems where the basic state variables do not possess a definite time reversal symmetry [6, 7]. However, in spite of the fact that the theory of fluctuations for nonequilibrium fluids was initiated in the late 1970s, and was pursued by many authors [8], still nowadays several questions concerning the nature of hydrodynamic fluctuations in stationary nonequilibrium states (*NESS*) are of current active interest. One of these issues is the long-range character of these fluctuations, especially far away from instability points [9]. Thermal fluctuations in an equilibrium fluid always give rise to short-range equal time correlation functions, except close to a critical point. But when external gradients are applied, equal-time correlation functions can develop long-range contributions, whose nature is very different from those in equilibrium. For many models and systems in nonequilibrium states, it has been shown theoretically that the existence of the so-called generic scale invariance is the origin of the long-range nature of the correlation functions [10, 11].

In the case of a simple fluid in a thermal gradient, the structure factor, which determines the intensity of the Rayleigh scattering, diverges as k^{-4} for small values of the wave number k. This amounts to an algebraic decay of the density-density correlation function, a feature that has been verified experimentally [12–14]. However, there are few similar studies for *NESS* of complex fluids. Among these, the enhancement of concentration fluctuations in polymer solutions under external hydrodynamic and electric fields [15], or the case of a polymer solution subjected to a stationary temperature gradient in the absence of any flow [16], has been discussed. Also, the behavior of fluctuations about some *NESS* has been analyzed in the case of thermotropic nematic liquid crystals. Specific examples are the nonequilibrium situations generated by a static temperature gradient [17], a stationary shear flow [18] or by an externally imposed constant pressure gradient [19, 20]. In the first two cases, it was found that the nonequilibrium contributions to the corresponding light scattering spectrum were small, but in the case of a Poiseuille flow induced by an external pressure gradient, the effect may be quite large. To our knowledge, however, at present, there is no experimental confirmation of these effects, in spite of the fact that for nematics, the scattered intensity is several orders of magnitude larger than for ordinary simple fluids.

When a hydrodynamic system relaxes from a state of thermodynamic equilibrium to another, almost all its degrees of freedom will return to that equilibrium value in a short, finite time τ determined by the microscopic interactions of the system. There are, however, some other degrees of freedom of collective character, the hydrodynamic modes, which will decay much more slowly. When $\tau \to \infty$, its characteristic frequencies $\omega \to 0$ ($\omega \sim 1/\tau$), when $k \to 0$. Such is the case, for example, of the propagation of sound waves and the conduction of heat in a simple fluid [21]. Hydrodynamics allows to describe these modes or degrees of freedom of greater duration, through the laws of conservation and balance of the system, and, as in the case of ordered systems, by the continuous breaking of symmetries [22, 23].

The central purpose of this work is to briefly review the general procedure developed by Fox and Uhlenbeck and show that it may be employed to treat fluctuating complex fluid systems like a thermotropic nematic liquid crystal (*NLC*) in a *NESS*. In particular, we describe the dynamics of the fluctuations of its hydrodynamic variables induced by a stationary temperature gradient and under the influence of gravity (a Rayleigh-Bénard system) on a nematic layer confined between two parallel horizontal plates in a steady state in a nonconvective regime [24–26]. Once the dynamics of fluctuation is established, we calculate the time-dependent correlation functions in equilibrium between the fluctuating hydrodynamic variables, quantities that allow to obtain the transport properties of the system [27, 28]. One of these properties is the dynamic structure factor $S\left(\vec{k}, \omega\right)$ of the system, which measures the magnitude of the changes in energy and momentum between the light beam and the fluid as functions of the wave vector $\vec{k}$ and ω.

For simple fluids with fixed $\vec{k}$, $S\left(\vec{k}, \omega\right)$ consists of three well-separated Lorentzian features: a line or central peak (Rayleigh peak) located at $\omega = 0$ and two Brillouin peaks symmetrically located with respect to the central one [29, 30]. These three lines are directly related to the hydrodynamic modes of the simple fluid, and from them, it is possible to obtain relevant information about transport properties. For instance, the Rayleigh line, associated with a thermal diffusive mode, is due to the fluctuations of the entropy (or temperature) that diffuse in the fluid and its width is proportional to the thermal diffusivity. On the other hand, the Brillouin lines are related to two acoustic propagative modes and are the result of the coupled dynamics of the pressure fluctuations and a component of the flow velocity that are transmitted with the speed of sound in the medium. Their widths are proportional to the absorption of sound.

In the case of an anisotropic system like a *NLC*, fluctuating hydrodynamic theories have recently been proposed [22, 31] based on the methodology proposed by Landau and Lifshitz [1]. However, this analysis of the fluctuations of the nematic hydrodynamic variables is not precise, since it does not take into account the parity with respect to time reversal, so their description using the Onsager-Machlup formalism would be strictly inadequate. The correct theoretical framework should be the more general theory of Fox and Uhlenbeck [3–5, 20, 24]. However, although a *NLC* disperses light by several orders of magnitude more than an ordinary fluid [32], from both a theoretical and experimental point of view, the studies corresponding to the behavior of the fluctuations in these media around stationary states out of equilibrium are rather scarce. From the theoretical point of view, and only for the case of the transverse hydrodynamic variables [33], some studies of the behavior of orientational fluctuations have been carried out when analyzing the effect produced in the light scattering spectrum of a *NLC* in *NESS* induced by the presence of uniform temperature gradients [17] and by the action of a shear flow [18]. In both cases, it has been found that the effect of fluctuations in the light scattering spectrum is small, being difficult to detect experimentally. On the other hand, as far as we know, no theoretical study has been carried out on the behavior of the longitudinal variables of a nematic and much less on its spectrum of light scattering, both in states of thermodynamic equilibrium and outside of it. This is an open research topic. Nor have been performed analyzes of stationary states generated by other types of external gradients in these systems, with which could be obtained qualitatively and quantitatively much greater effects than those reported so far in the literature for simple fluids. It should be mentioned that although preliminary attempts have been made to calculate the transverse hydrodynamic modes of a nematic [34, 35], there are few studies that also involve the corresponding longitudinal modes [31]. Unfortunately, a clear and systematic method to derive the set of complete, transverse, and longitudinal hydrodynamic modes of a *NLC* is still lacking.

By introducing an alternative set of state variables that takes into account the asymmetry presented by both, the velocity and the director fields due by their mutual coupling, two groups of fluctuating variables, namely, longitudinal and transverse, can be clearly identified. Both set of variables are completely decoupled: there are five in the first and two in the second group. The longitudinal variables in turn can be separated into two mutually independent sets. The first is composed of two variables whose dynamics determine the existence of acoustic propagation modes; while the second, formed by three variables, giving rise to three hydrodynamic modes: one related to the orientation of the director and two more, the so-called visco-heat modes, that result from the coupling of the thermal diffusive and shear modes due by the presence of the gradient thermal and the gravitational field. As will be discussed later on, from the set of transverse variables, two hydrodynamic modes emerge: one due to the orientation of the director and another one due to shearing. Altogether, there are seven nematic hydrodynamic modes: five longitudinal and two transversal. As will be shown below, the applied gradient of temperature and gravitational field produce their greatest effect in the pair of visco-heat modes, which is quantified in them by means of the Rayleigh quotient R/R_c, where R is the number of Rayleigh and R_c the value that it reaches when in the nematic initiates the convection.

2. Liquid crystalline phases

The liquid crystal phase is a well-defined and specific phase of matter characterized by a remarkable anisotropy in many of their physical properties as solid crystals do, although they are able to flow. Liquid crystal phases that undergo a phase transition as a function of temperature (thermotropics) exist in relatively small intervals of temperature lying between solid crystals and isotropic liquids. Due to this intermediate nature, sometimes, these states are called also mesophases [32]. In general, liquid crystals are synthesized from organic molecules, some of which are elongated and uniaxial, so they can be represented as rigid rods; others are formed by disc-like molecules [35]. This molecular anisotropy is manifested macroscopically through the anisotropy of the mechanical, optical, and transport properties of these substances. The typical dimensions of the lengths of this type of structures are some tens of angstroms.

Liquid crystals are classified by symmetry. As it is well known, isotropic liquids with spherically symmetric molecules are invariant under rotational, $O(3)$, and

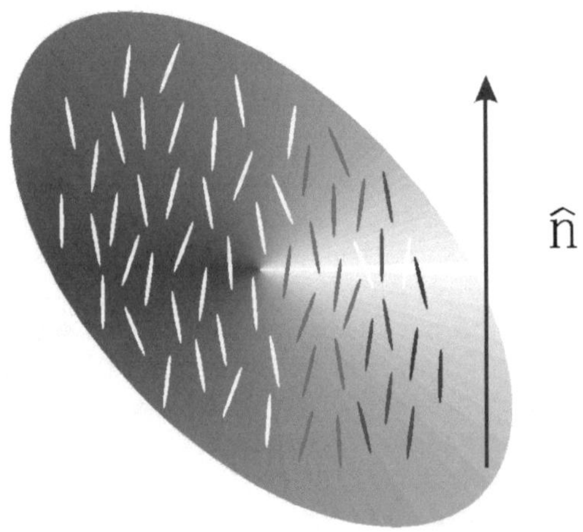

Figure 1.
Representation of the average orientation of the molecules of a thermotropic NLC by means of the director vector $\hat{n}$.

translational, $T(3)$, transformations. Thus, the group of symmetries of an isotropic liquid is $O(3) \times T(3)$. However, by decreasing the temperature of these liquids, the translational symmetry $T(3)$ is usually broken corresponding to the isotropic liquid-solid transition. In contrast, for a liquid formed by anisotropic molecules, by diminishing the temperature, the rotational symmetry $O(3)$ is broken, which leads to the appearance of a liquid crystal. The mesophases for which only the rotational invariance has been broken are called nematics. As shown, the centers of mass of the molecules of a nematic have arbitrary positions, whereas the principal axes of their molecules are spontaneously oriented along a preferred direction. If the temperature decreases even more, the symmetry $T(3)$ is also partially broken. The mesophases exhibiting the translational symmetry $T(2)$ are called smectics [36].

This preferential direction is described by a local unitary vector field, $\hat{n}$, called the director. This vector is easily distorted by the presence of electric and magnetic fields, as well as by the surfaces of the containers of the liquid crystals if they have been prepared properly [32]. With respect to *NLC*, it is important to point out that the director's orientation does not distinguish between the $\hat{n}$ and $-\hat{n}$ directions (nematic symmetry). A schematic representation of the order presented by the molecules in a nematic is shown in **Figure 1** .

3. Model

Consider a *NLC* thin layer of thickness d under the presence of a constant gravitational field $\vec{g} = -g\hat{z}$, where g denotes its magnitude and $\hat{z}$ the unit vector along the z axis. The initial configuration of the layer is homeotropic with a preferential orientation $\hat{n}_0$ along the z axis, as depicted in **Figure 2**. The nematic is confined between two parallel flat plates kept at fixed temperatures T_1 and T_2 ($T_1 < T_2$), so that a uniform temperature gradient $\nabla_z T \equiv -\alpha\hat{z}$ is established downward in the layer. The situation where the temperature gradient goes from bottom to top can also be considered, and in this case, $\nabla_z T \equiv \alpha\hat{z}$. The gravitational force induces a pressure gradient, $\nabla_z p = -\rho g\hat{z}$, where ρ is the mass density.

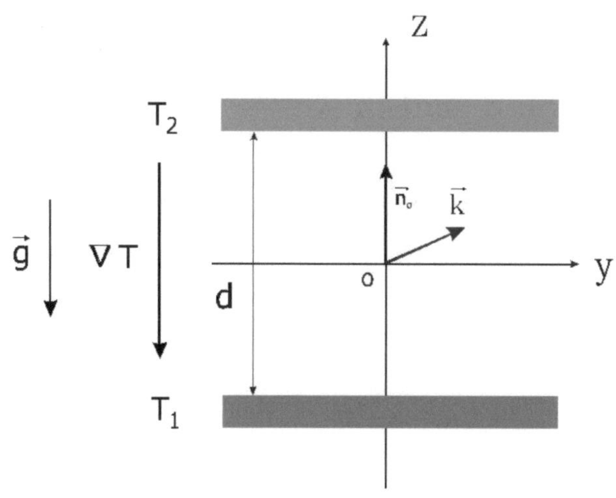

Figure 2.
The NLC cell subject to a constant gravitational field $\vec{g}$ and an external uniform temperature gradient ∇T. $\vec{k}$ is the scattering vector.

3.1 Stationary state

The external gradients drive the nematic layer into a nonequilibrium steady state. We shall assume that the temperature difference $T_1 - T_2$ amounts only to a few degrees, so that there are no nematic layer flows $\left(v_i^{st} = 0\right)$, nor instabilities of the Rayleigh-Bénard type. In this *NESS,* we choose as the nematodynamic variables the set $\Psi = \{\rho, s, v_i, n_i\}$, where $s\left(\vec{r}, t\right)$ is the specific entropy density (entropy per unit mass), the hydrodynamic velocity is $v_i\left(\vec{r}, t\right)$ and $n_i\left(\vec{r}, t\right)$ is the director field. It is to be expected that in this steady state, the changes in Ψ^{st} will only occur in the z direction, so that $\Psi^{st} = \Psi[p(z), T(z)]$, where p is the local pressure. We assume that Ψ^{st} admits an expansion of the Taylor series around an equilibrium state $\left(T_0, p_0\right)$ at $z_0 = 0$, and we consider only first-order terms in the gradients. Thus, by setting the values of the temperature at the plates, $T_1 = T(z = -d/2)$ and $T_2 = T(z = d/2)$, the steady temperature profile is completely determined by:

$$T^{st} = T(z) = T_0 + \frac{dT}{dz}z = T_0\left(1 - \frac{\alpha}{T_0}z\right), \tag{1}$$

where $T_0 \equiv T^{st}(z = 0) = (T_1 + T_2)/d$ and $\alpha \equiv \Delta T/d$, with $\Delta T \equiv T_1 - T_2$. In what follows, we shall only consider $T_0 \approx 3 \times 10^2 K$, and it will be convenient to introduce the *effective temperature gradient* $\nabla_z T^{st} \equiv X\hat{z}$ as [37],

$$X \equiv -\alpha + \frac{g\beta T_0}{c_p}, \tag{2}$$

which contains explicitly the contributions of both external forces. In Eq. (2), c_p is the specific heat at constant pressure, β is the thermal expansion coefficient, which satisfies the relationship $\beta^2 \equiv (\gamma - 1)c_p/T_0 c_s^2$, where c_s is the adiabatic sound velocity in the nematic, $\gamma \equiv c_p/c_v = c_s^2/c_T^2$, being c_v the specific heat at constant volume and c_T the isothermic sound velocity in the nematic.

4. Nematodynamic equations

The geometry of the proposed model allows us to separate the hydrodynamic variables into transverse (T) and longitudinal (L) variables with respect to $\hat{n}_0$ and $\vec{k}$, [33]. The former set is $\Psi^T\left(\vec{r}, t\right) \equiv \{v_x, n_x\}$, while the latter reads $\Psi^L\left(\vec{r}, t\right) \equiv \{p, v_y, v_z, s, n_y\}$. We want to describe the stochastic dynamics of the spontaneous thermal deviations (fluctuations) $\delta\Psi\left(\vec{r}, t\right) = \Psi\left(\vec{r}, t\right) - \Psi^{st}$ around the above defined stationary state. A complete set of stochastic equations for the space-time evolution of the fluctuations is obtained by linearizing the general nematodynamic equations [20, 22, 24], and by using the *FH* formalism. This starting set of equations is given explicitly by Eqs. (19)–(22) in Ref. [25]. However, since for the nematic mesophase, the rotational invariance has been broken, it is convenient to rewrite these nematodynamic equations in a representation which takes into account that a symmetry breaking has occurred along the z axis.

In order to take into account the effect of the intrinsic anisotropy of the fluid in the dynamics of the fluctuations, as well as to facilitate the calculation of the nematic modes and the spectrum of light scattering, it is convenient to introduce a new state variables. In the case of the present model, owing to the initial orientation of the director $\hat{n}_i^{st}$, the *NLC* exhibits several symmetries: rotational invariances around the z axis, symmetry under inversions with respect to both, the xy plane and with respect to reflections on planes containing the z axis. A proper set of variables for this purpose was proposed long ago [38, 39], in terms of the variables $\{\delta p, \delta\varphi, \delta s, \delta\xi, \delta f_1, \delta\psi, \delta f_2\}$, defined in detail in Eqs. (6)–(10) in Ref. [26] (or Eqs. (53)–(57) in [25]). In this new representation, the complete set of stochastic hydrodynamic equations for the fluctuations takes an alternative form given by Eqs. (11)–(17) in Ref. [26] (or Eqs. (58)–(64) in [25]). The matrix representation of the Fourier transformation of this set of equations is given by:

$$\frac{\partial}{\partial t}\delta \vec{X}\left(\vec{k},t\right) = -M\delta \vec{X}\left(\vec{k},t\right) + \vec{\Theta}\left(\vec{k},t\right), \tag{3}$$

where $\delta \vec{X}\left(\vec{k},t\right) = \left(\delta\vec{X}^L, \delta\vec{X}^T\right)^t$ with $\delta\vec{X}^L\left(\vec{k},t\right) = \left(\delta\tilde{p}, \delta\tilde{\varphi}, \delta\tilde{s}, \delta\tilde{\xi}, \delta\tilde{f}_1\right)^t$ and $\delta\vec{X}^T\left(\vec{k},t\right) = \left(\delta\tilde{\psi}, \delta\tilde{f}_2\right)^t$. The superscript t denotes the transpose, while L and T indicate, respectively, the longitudinal and transverse sets of variables. In Eq. (3), M stands for a 7×7 hydrodynamic matrix which is diagonal in the 5×5 N^L and the 2×2 N^T blocks. The explicit form of these matrices is not necessary in our discussion; however, they are given explicitly by Eqs. (21) and (22) in Ref. [26] (see also Eqs. (72) and (73) in Ref. [25]). The stochastic terms, $\vec{\Theta}\left(\vec{k},t\right)$, in Eq. (3) are given by the column vector $\vec{\Theta}\left(\vec{k},t\right) = \left(\vec{\Theta}^L, \vec{\Theta}^T\right)^t$ which explicit form of its components can be found in Eqs. (32) and (33) in Ref. [26] (or Eqs. (84) and (85) of Ref. [25]). It is important to emphasize that as a consequence of this change of representation, in this last system, it can be clearly seen how the nematic variables are separated in two sets completely independent: the five longitudinal $\left\{\delta\tilde{p}, \delta\tilde{\varphi}, \delta\tilde{s}, \delta\tilde{\xi}, \delta\tilde{f}_1\right\}$ and the two transverse $\left\{\delta\tilde{\psi}, \delta\tilde{f}_2\right\}$.

However, in order to facilitate the calculation of the hydrodynamic modes, we define a new set of variables having the same dimensionality, $\left[\delta z_j\left(\vec{k},t\right)\right] = M^{1/2}L^{-1/2}t$ $(j = 1,\dots, 7)$: $z_1 \equiv \left(\rho_0 c_s^2\right)^{-1/2}\delta\tilde{p}$, $z_2 \equiv \left(\rho_0 k^{-2}\right)^{1/2}\delta\tilde{\varphi}$, $z_3 \equiv \left(\rho_0 T_0 c_p^{-1}\right)^{1/2}\delta\tilde{s}$, $z_4 = \left(\rho_0 k^{-4}\right)^{1/2}\delta\tilde{\xi}$, $z_5 \equiv \left(\rho_0 c_s^2 k^{-2}\right)^{1/2}\delta\tilde{f}_1$, $z_6 \equiv \left(\rho_0 k^{-2}\right)^{1/2}\delta\tilde{\psi}$, $z_7 \equiv \left(\rho_0 c_s^2 k^{-2}\right)^{1/2}\delta\tilde{f}_2$. In terms of these new variables, the system of equations (3) is rewritten in the more compact form as:

$$\frac{\partial}{\partial t}\vec{Z}\left(\vec{k},t\right) = -N\vec{Z}\left(\vec{k},t\right) + \vec{\Xi}\left(\vec{k},t\right), \tag{4}$$

where $\vec{Z}\left(\vec{k},t\right) = \left(\vec{Z}^L, \vec{Z}^T\right)^t$ with $\vec{Z}^L\left(\vec{k},t\right) = (z_1, z_2, z_3, z_4, z_5)^t$ and $\vec{Z}^T\left(\vec{k},t\right) = (z_6, z_7)^t$. In Eq. (4), N stands for a 7×7 hydrodynamic matrix which is diagonal in the 5×5 N^L and the 2×2 N^T blocks. Again, the explicit form of these matrices is not necessary in our discussion, but they are given explicitly by Eqs. (39)–(41) in Ref. [26] (see also Eqs. (94)–(96) in [25]). In Eq. (4), $\vec{\Xi}\left(\vec{k},t\right) = \left(\vec{\Xi}^L, \vec{\Xi}^T\right)^t$ is the stochastic term, composed by the longitudinal $\vec{\Xi}^L\left(\vec{k},t\right) = (\zeta_1, \zeta_2, \zeta_3, \zeta_4, \zeta_5)^t$ and transverse $\vec{\Xi}^T\left(\vec{k},t\right) = (\zeta_6, \zeta_7)^t$ noise vectors. The explicit form of the components ζ_m, $m = 1...7$, as well as their fluctuation-dissipation relations (*FDR*), can be found in Eqs. (169)–(175) and Eqs. (176)–(186), respectively, in Appendix A of [25].

5. Hydrodynamic modes

In order to find the hydrodynamic modes, or decay rates [37], we need the Fourier transform of the linear system (4), which yields an algebraic system of equations in terms of the variables $\vec{k}$ and ω. The hydrodynamic modes are obtained by calculating its eigenvalues $\lambda = i\omega$, given by the roots of the characteristic equation $p(\lambda) = p^L(\lambda)p^T(\lambda) = 0$, where $p^L(\lambda)$ and $p^T(\lambda)$ are the characteristic polynomials of fifth and second order in λ of the matrices N^L and N^T, respectively. These roots are calculated below.

5.1 Longitudinal modes

Following the method proposed by [13] for a simple fluid, it can be shown that longitudinal variables can be separated in turn and within a very good approximation, into two completely independent sets of variables, $\vec{Z}_X^L = (z_1, z_2)^t$ and $\vec{Z}_Y^L = (z_3, z_4, z_5)^t$, as it is shown in the Subsection 3.1 of Ref. [25], or in more detail in [24]. This approximation allows us to rewrite the characteristic polynomial of longitudinal variables as $p^L(\lambda) = p^L_{XX}(\lambda)p^L_{YY}(\lambda)$. It should be mentioned that $p^L_{XX}(\lambda)$ and $p^L_{YY}(\lambda)$ are polynomials of second and third degree in λ, and explicitly are given by the Eqs. (44) and (45) in Ref. [26] (or Eqs. (117) and (118) in [25]).

While there is no analytical difficulty to solve the quadratic and cubic equations $p^L_{XX}(\lambda)$ and $p^L_{YY}(\lambda)$, the explicit form of their exact roots can be quite complicated. However, it is possible to estimate them following a procedure based partially on a method suggested in Ref. [40], which allows to identify the following quantities in the equation for $p^L_{YY}(\lambda)$, namely, $(\gamma - 1)D_T k^2$, $\sigma_1 k^2$, $k^2 c_s^2$ and $g^2 k_\parallel^2/(c_s^2 k^2)$. They depend on the anisotropic coefficients of diffusivity D_T and on the viscosity σ_1. The former quantity is a function of the parallel $\chi_\parallel$ and perpendicular $\chi_\perp$ components of thermal diffusivity, while the latter depends on the nematic viscosity coefficients ν_i $(i = 1, ..., 5)$ (see Eqs. (23) and (24) in Ref. [26], or Eqs. (74) and (75) in [25]). In the same way, in the equation for $p^L_{YY}(\lambda)$, the following quantities can be identified, $g\alpha\beta\frac{k_\perp^2}{k^2}$, $gX\beta\frac{k_\perp^2}{k^2}$, $D_T k^2$, $\Omega\chi_a k^2$, $\sigma_3 k^2$, $\frac{K_1}{\gamma_1}k^2$, $\frac{\Omega^2 K_I}{\rho_0}k^4$, which depend on the anisotropic coefficients of viscosity σ_3, of elasticity K_I, symmetry Ω (see, respectively, Eqs. (26), (28), and (30) in [26]), as well as the anisotropy $\chi_a = \chi_\parallel - \chi_\perp$ and the torsional viscous coefficient γ_1. We now compare all these quantities with $\overline{\omega} \equiv c_s k$, by introducing the (small) reduced dimensionless quantities:

$$a_0 \equiv \frac{g\alpha\beta}{\overline{\omega}^2}\frac{k_\perp^2}{k^2}, \quad a_0' \equiv \frac{gX\beta}{\overline{\omega}^2}\frac{k_\perp^2}{k^2}, \quad a_0'' \equiv \frac{g^2 k_\parallel^2}{\overline{\omega}^2 c_s^2 k^2}, \quad a_1 \equiv \frac{D_T k^2}{\overline{\omega}}, \quad a_1' \equiv \frac{\Omega\chi_a k^2}{\overline{\omega}},$$
$$a_2 \equiv \frac{\sigma_1 k^2}{\overline{\omega}}, \quad a_3 \equiv \frac{\sigma_3 k^2}{\overline{\omega}}, \quad a_5 \equiv \frac{K_I k^2}{\gamma_1 \overline{\omega}}, \quad a_6 \equiv \frac{\Omega^2 K_I k^4}{\rho_0 \overline{\omega}^2}. \tag{5}$$

The relevant point for our purpose is to realize that for most nematics at ambient temperatures, ρ_0 and Ω are of order of magnitude 1, $\gamma_1 \sim 10^{-1}$, χ_i and ν_i are of order 10^{-2}–10^{-3}, $K_i \sim 10^{-6}$–10^{-7}, while $\beta \sim 10^{-4}$ [32]. If we consider that $\alpha \lesssim 1$ and

$g \sim 10^3$, and knowing that in a typical light scattering experiments $k = 10^5 \text{cm}^{-1}$ and $c_s = 1.5 \times 10^5 \text{cms}^{-1}$ [41], the quantities given in Eq. (5) have the following orders of magnitude: $a_0 \sim 10^{-21}$, $a_0' \sim 10^{-21}$, $a_0'' \sim 10^{-24}$, $a_1 \sim 10^{-3}$, $a_1' \sim 10^{-3}$, $a_2 \sim 10^{-2}$, $a_3 \sim 10^{-2}$, $a_5 \sim 10^{-5}$ and $a_6 \sim 10^{-6}$. We now follow the method of Ref. [40] and the solutions of the polynomial $p^L_{YY}(\lambda)$ may be obtained by a perturbation approximation in terms of these small quantities. However, in what follows, we improve this approximation by using its exact roots and by expressing them in terms of the reduced quantities (Eq. (5)) of order k^2 [24].

5.1.1 Sound longitudinal modes

They are the roots of the characteristic equation $p^L_{XX}(\lambda) = 0$. Its roots are complex conjugate and are given by (see Eqs. (47) and (48) in [26], or Eqs. (128) and (129) in [25]):

$$\lambda_1 \simeq \Gamma k^2 + ic_s k, \quad \lambda_2 \simeq \Gamma k^2 - ic_s k, \tag{6}$$

where $\Gamma \equiv \frac{1}{2}[(\gamma - 1)D_T + \sigma_1]$ is the anisotropic sound attenuation coefficient of the *NLC*. This result shows that the sound propagation modes, λ_1 and λ_2, are in complete agreement with those already reported in the literature for *NLC* [31, 34].

5.1.2 Visco-heat and director longitudinal modes

These modes are the roots of the characteristic equation $p^L_{YY}(\lambda) = 0$. In Ref. [26] (or in [25]), it is shown that, up to first order in the small quantities (Eq. (5)), these roots can be written approximately as:

$$\begin{aligned} \lambda_{3,4} &= \frac{1}{2}\left(D_T k^2 + \sigma_3 k^2 - \frac{\Omega^2 K_I k^4}{\rho_0 \sigma_3 k^2}\right) \\ &\mp \frac{1}{2}\sqrt{\left(D_T k^2 + \sigma_3 k^2 - \frac{\Omega^2 K_I k^4}{\rho_0 \sigma_3 k^2}\right)^2 - 4 D_T k^2 \sigma_3 k^2 \left(1 - \frac{R}{R_c}\right)}, \end{aligned} \tag{7}$$

and

$$\lambda_5 \simeq \frac{K_I k^2}{\gamma_1} + \frac{\Omega^2 K_I k^4}{\rho_0 \sigma_3 k^2}, \tag{8}$$

with

$$\frac{R\left(\vec{k}\right)}{R_c} \equiv -\frac{g\beta \hat{k}_\perp^2}{D_T \sigma_3 k^4}\left[X + \frac{\alpha \Omega \chi_a}{D_T \sigma_3}(\sigma_3 + D_T)\right], \tag{9}$$

where $\hat{k}_\perp^2 \equiv k_\perp^2 / k^2$. In Eq. (7), $R \equiv \frac{\beta g \Delta T d^3}{\sigma_3 \chi}$ is the Rayleigh number and R_c denotes its critical value above which convection sets in. It should be emphasized that our results are expressed in terms of the ratio $R\left(\vec{k}\right)/R_c$ and are, therefore, independent of the value of the separation d between the plates. However, the appropriate value of d in an experiment should be chosen with an experimental criterion [42].

The Rayleigh-number ratio $R\left(\vec{k}\right)/R_c$ contains two contributions: the first term is due to the presence of the effective temperature gradient X, which depends on both, the temperature gradient α and the gravity field g. The second term is entirely a contribution due to α and the nematic anisotropy χ_a. For typical nematics and conventional light scattering experiments, both contributions are approximately of order 10^{-16}.

The decay rates λ_3 and λ_4 for an inhomogeneous nematic given by Eq. (7) are called visco-heat modes, because they are composed of the coupling between the thermal $D_T k^2$ and shear $\sigma_3 k^2 - \frac{\Omega^2 K_I k^4}{\rho_0 \sigma_3 k^2}$ diffusive modes through the ratio $R\left(\vec{k}\right)/R_c$. The nature of these modes may be propagative or diffuse, as will be shown below.

5.1.3 Values of $R\left(\vec{k}\right)/R_c$

The three nematic modes (7) and (8) could be two propagative and one diffusive, or all of them completely diffusive; its nature depends on the values assumed by the ratio $R\left(\vec{k}\right)/R_c$. For simple fluids, these features have been predicted theoretically and corroborated experimentally, but to our knowledge, not for an *NLC*. In this sense, the following results suggest that it might be feasible to be also verified experimentally for nematics.

5.1.3.1 Propagative and diffusive modes

If we take into account the orders of magnitude of the small quantities (Eq. (5)), the nematic modes (7) and (8) in general are real and different. Nevertheless, it may happen that these modes may be transformed into one real and two complex conjugate roots. This occurs if $R\left(\vec{k}\right)/R_c < R_0$, where

$$R_0 \equiv -\frac{\left[\left(\sigma_3 - \frac{\Omega^2 K_I}{\rho_0 \sigma_3}\right) - D_T\right]^2}{4 D_T \sigma_3}, \tag{10}$$

which is always negative. Thus, if we consider the orders of magnitude of the involved quantities and typical light scattering experiment values of k, $D_T k^2 \sim 10^7$, $\sigma_3 k^2 \sim 10^8$, and $\frac{\Omega^2 K_I k^4}{\rho_0} \sim 10^{14}$, then $R_0 \cong -10^1$ and the visco-heat modes, Eq. (7), will be propagative when $R\left(\vec{k}\right)/R_c \lesssim -10^1$. This situation corresponds to the propagation region indicated in **Figure 3**. The decay rate λ_5, Eq. (8), remains to be real. It is worth emphasizing that this case corresponds to overstabilized states, where out of the three decay rates, two are propagative visco-heat modes and the other one is completely diffusive. According to Eq. (9), this occurs if the α contained in the effective temperature gradient X changes its sign and increases by several orders of magnitude, situation that may be achieved by reversing the direction in which the temperature gradient is applied, i. e., when heating from below, and by increasing its intensity. As far as we know, there are no theoretical analyses nor experimental evidence for the existence of visco-heat propagating modes in *NLC* under the presence of a temperature gradient and a uniform gravitational field. Given that in simple fluids, under these conditions, there are analytical [8, 37, 38] and experimental [43] studies that support the presence of visco-heat

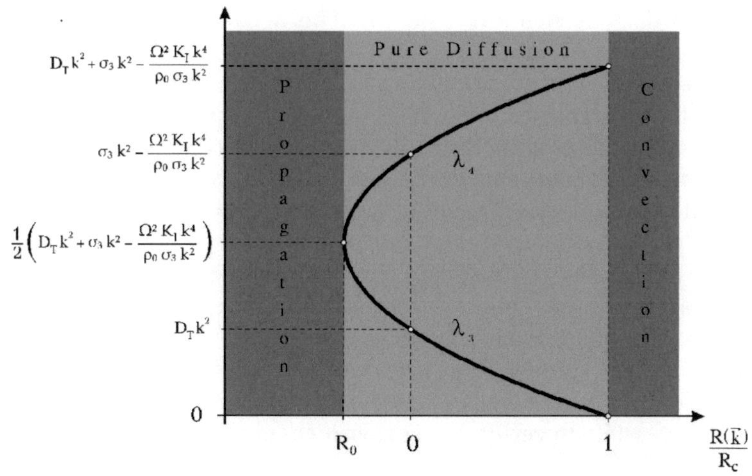

Figure 3.
The real part of the nematic visco-heat modes λ_3 and λ_4 as a function of the Rayleigh ratio $R(\vec{k})/R_c$. When $R(\vec{k})/R_c < R_0$, both modes are propagative; if $R_0 \leq R(\vec{k})/R_c \leq 1$, both are completely diffusive. For $R(\vec{k})/R_c = R_0$, both modes are equal. In equilibrium, $R(\vec{k})/R_c = 0$, and the onset of convection occurs for $R(\vec{k})/R_c = 1$.

propagation modes, this prediction suggests that it may be worth to design experiments to corroborate this phenomenon for nematics.

5.1.3.2 Pure diffusive modes

When $R_0 \leq R(\vec{k})/R_c \leq 1$, the two visco-heat modes preserve the same form as in Eq. (7) and the other one remains identical to Eq. (8), but all are real and completely diffusive. In this regime, the following cases are of special interest. For instance, if $R(\vec{k})/R_c = R_0$, then the visco-heat modes (7) reach the same value, and consequently, the three decay rates are:

$$\lambda_{3,4} = \frac{1}{2}\left(D_T k^2 + \sigma_3 k^2 - \frac{\Omega^2 K_I k^4}{\rho_0 \sigma_3 k^2}\right), \tag{11}$$

and λ_5, that takes the same form as in Eq. (8). These visco-heat modes are identified at the vertex of the parabola in **Figure 3**.

Since for nematics, $\sigma_3 - \frac{\Omega^2 K_I}{\rho_0 \sigma_3}$ is usually greater than D_T, it can be seen from Eq. (7) that, as $R(\tilde{k})/R_c$ grows and approaches 1, the magnitude of the heat diffusive mode decreases, whereas the one of the shear mode increases. At the onset of convections regime, $R(\vec{k})/R_c = 1$, i. e., when R reaches its critical value R_c and the two visco-heat modes (7) are simplified to:

$$\lambda_3 = 0, \tag{12}$$

$$\lambda_4 = D_T k^2 + \sigma_3 k^2 - \frac{\Omega^2 K_I k^4}{\rho_0 \sigma_3 k^2}, \tag{13}$$

while the third, λ_5, is identical to Eq. (8). This behavior for the decay rates λ_3 and λ_4 is also shown in **Figure 3**.

It should be noted that our expressions for these three decay rates are not in agreement with those reported for an *NLC* in the literature [44, 45]. In these works, the director mode tends to zero, the shear mode does not change and there is an additional mode which is the sum of the thermal and director modes. In contrast, we have found that the thermal mode λ_3 vanishes, the director mode λ_5 is virtually unchanged, while λ_4 has contributions from the thermal and shear diffusive modes. We know that this phenomenon also occurs in the simple fluid, where there are two diffusive modes, the thermal mode also vanishes and the other one has contributions from the shear and thermal modes. In other words, our results reduce to the corresponding one for a simple fluid as R reaches its critical value R_c. Because for a simple fluid, these features have been predicted theoretically, our results suggest that it might be feasible to verify them experimentally also for nematics [8, 37, 38].

5.2 Transverse modes

As mentioned earlier, $p^T(\lambda)$ is the characteristic polynomial of second order in λ of the matrix N^T. The corresponding transverse hydrodynamic modes are the roots of this equation $p^T(\lambda) = 0$.

5.2.1 Shear and director transverse modes

Accordingly, the shear and director transverse modes are the roots of $p^T(\lambda) = 0$, and are given by Eq. (63) in Ref. [26] (or by Eq. (157) in [25]). Following again the approximate method of small quantities used previously, the quantities σ_4, $K_{II}k^2/\gamma_1$ and $\lambda_+ K_{II}k^2k_\parallel^2/\rho_0$, may be identified in this equation. In terms of them, we have another set of anisotropic coefficients given by the viscosity σ_4, the elasticity K_{II}, and symmetry λ_+ (see, respectively, Eqs. (27), (29), and (31) in [26]). We also define the small or reduced dimensionless quantities, analogous to those defined in Eq. (5), namely, $a_4 \equiv \frac{\sigma_4 k^2}{\overline{\omega}}$, $a_5' \equiv \frac{K_{II}k^2}{\gamma_1 \overline{\omega}}$, $a_6' \equiv \frac{\lambda_+^2 K_{II}}{\rho_0 \overline{\omega}^2} k^2 k_\parallel^2$, where again $\overline{\omega} \equiv c_s k$. It should be noted that the viscous coefficient σ_4 only depends on the viscous coefficients ν_2, ν_3, while the elastic coefficient K_{II} depends on the two Frank elastic constants K_2 and K_3. Since for typical nematics $\lambda_+ \sim 1$, $\gamma_1 \sim 10^{-1}$, $\sigma_4 \sim 10^{-2}$, $K_{II} \sim 10^{-6}$ [32], and also by taking into account that $c_s \sim 10^5$, $k \sim 10^5$, $g \sim 10^3$, the quantities a_4, a_5' and a_6' have the orders of magnitude $a_4 \sim 10^{-2}$, $a_5' \sim 10^{-5}$, and $a_6' \sim 10^{-6}$. According to Eqs. (64) and (65) in Ref. [26] (or Eqs. (167) and (168) in [25]), up to first order in such small amounts, these two roots can be written as:

$$\lambda_6 = \sigma_4 k^2 - \frac{\lambda_+^2 K_{II} k^2 k_\parallel^2}{\rho_0 \sigma_4 k^2}, \qquad \lambda_7 = \frac{K_{II} k^2}{\gamma_1} + \frac{\lambda_+^2 K_{II} k^2 k_\parallel^2}{\rho_0 \sigma_4 k^2} \tag{14}$$

It should be noted that these shear and director diffusive transverse modes also match completely with those already reported for nematics [22, 31, 32].

6. The equilibrium and simple fluid limits

From the hydrodynamic modes calculated for an *NLC* in a *NESS* determined by a Rayleigh-Bénard system, it is possible to obtain, as limit cases, the corresponding modes of a nematic in the state of equilibrium and those of a simple fluid under the same nonequilibrium regime. Both situations are of physical interest and are discussed below.

6.1 Nematic in equilibrium

It has been found that for an *NLC* in a *NESS*, the effects of the external gradients α and g are only manifested in the coupling of the thermal diffusive and shear longitudinal modes, which gives rise to the visco-heat modes $\lambda_{3,4}$ indicated, respectively, by means of Eq. (7). If the nematic layer is in a state of homogeneous thermodynamic equilibrium, $g = 0$ and $\alpha = 0$, and therefore $X = 0$ and $R\left(\vec{k}\right)/R_c = 0$. Thus, the hydrodynamic modes of a nematic, in the state of equilibrium (denoted by the superscript e), are composed of five longitudinal and two transverse modes. The longitudinal modes are integrated by the two acoustic propagatives λ_1^e and λ_2^e given by Eq. (6); as well as by the three diffusives, which consist of one thermal:

$$\lambda_3^e = D_T k^2, \tag{15}$$

another of shear:

$$\lambda_4^e = \sigma_3 k^2 - \frac{\Omega^2 K_I k^2}{\rho_0 \sigma_3} \tag{16}$$

and one more of the director, λ_5^e, which is the same as Eq. (8). The longitudinal diffusive modes (15) and (16) are obtained precisely from Eq. (7), since in this, the Rayleigh ratio, given by Eq. (9), is zero if α and g vanish. Moreover, the pair of transverse modes consist of the shear and director modes λ_6^e and λ_7^e which are equal to the Eq. (14). It is necessary to mention that the decay rates λ_i^e $(i = 1...7)$ are well known in the literature [22, 31, 46]. Note that λ_3^e and λ_4^e are shown in the middle part of **Figure 3**.

6.2 Simple fluid in a Rayleigh-Bénard system

Given that in the isotropic limit (simple fluid limit), the degree of nematic order goes to zero, n_i is no longer a hydrodynamic variable, and the elastic constants K_i (for $i = 1, 2, 3$) and the kinetic parameters γ_1, λ vanish. Also, $\chi_\perp$ and $\chi_\parallel$ are reduced to the coefficient of thermal diffusivity χ and $\chi_a = 0$. On the other hand, the nematic viscosities are reduced in the following way: $\nu_1 \to \eta$, $\nu_2 \to \eta$, $\nu_3 \to \eta$, $\nu_4 \to \zeta + \frac{1}{3}\eta$, $\nu_5 \to -\frac{2}{3}\eta + \zeta$, where η and ζ denote, respectively, the shear and volumetric viscosities of the simple fluid. As a result, from Eqs. (23)–(31) in Ref. [26] (or Eqs. (74)–(82) in [25]), it follows that in the isotropic limit $D_T \to \chi$, $\sigma_1 \to \frac{1}{\rho_0}\left(\frac{4}{3}\eta + \zeta\right)$, $\sigma_2 \to 0$, $\sigma_3 \to \nu$, $\sigma_4 \to \nu$, where $\nu \equiv \eta/\rho_0$ is the kinematic viscosity, whereas $K_I \to 0$, $K_{II} \to 0$, and $\Omega \to 0$. Consequently, by making the identifications indicated above, the corresponding hydrodynamic modes of a simple fluid can be obtained when it is in a Rayleigh-Bénard system. Thus, according to Eq. (6), a simple fluid has the two acoustic propagative modes:

$$\lambda_1 \simeq \Gamma' k^2 + i c_s k, \quad \lambda_2 \simeq \Gamma' k^2 - i c_s k, \tag{17}$$

where c_s corresponds to the adiabatic velocity of the sound in this medium and $\Gamma' \equiv \frac{1}{2}\left[(\gamma-1)\chi + \frac{1}{\rho_0}\left(\frac{4}{3}\eta + \zeta\right)\right]$ is the corresponding coefficient of sound attenuation. On the other hand, according to the Eq. (7), the longitudinal visco-heat modes are:

$$\lambda_{3,4} \simeq \frac{1}{2}(\chi + \nu)k^2 \mp \frac{1}{2}\sqrt{(\chi+\nu)^2 k^4 - 4\chi\nu k^4\left(1 - \frac{R}{R_c}\right)}. \tag{18}$$

In the isotropic limit of the simple fluid, $\lambda_5 = \lambda_7 = 0$, so that, according to the Eq. (14), the only transverse mode of this substance in a Rayleigh-Bénard system is:

$$\lambda_6 = \nu k^2. \tag{19}$$

In Eq. (18), the ratio $R\left(\vec{k}\right)/R_c$ is defined as:

$$\frac{R\left(\vec{k}\right)}{R_c} \equiv -\frac{g\beta X \hat{k}_\perp^2}{\chi\nu k^4}, \tag{20}$$

which, in this limit case, can be derived from Eq. (9). It should be pointed out that Eq. (20) coincides with the Eq. (2.21) of reference [37]. The modes (17)–(19) are in complete concordance with those analytically calculated in [8, 37, 38].

Moreover, if in the coefficient matrix M of the stochastic system given by Eq. (20) in Ref. [26], the simple fluid limit is taken, it reduces to a matrix that is a generalization of the one given by the Eq. (6) in [38]. Additionally, if in the corresponding matrix M found for the simple fluid, the equilibrium limit is now considered, i. e., when α and g vanish, the resulting matrix is also reduced to that given by Eq. (4) of [38].

6.2.1 Values of $R\left(\vec{k}\right)/R_c$

The two visco-heat mode, as in the nematic, could be propagative or diffusive. These characteristics depend on the values assumed by the ratio $R\left(\vec{k}\right)/R_c$. For simple fluids, these have been predicted theoretically and corroborated experimentally.

6.2.1.1 Propagative modes

If $R\left(\vec{k}\right)/R_c < R_0$, where $R_0 \equiv -(\nu-\chi)^2/(4\chi\nu) < 0$, the visco-heat modes (18) will be propagative. According to Eq. (20), this occurs again if the α contained in X changes its sign and increases by several orders of magnitude, a situation that is achieved by inverting the temperature gradient (when heated from below and its intensity is increased). There are analytical [8, 37, 38] and experimental [43] studies that report, for simple fluids in these conditions, the presence of visco-heat propagative modes.

6.2.1.2 Pure diffusive modes

When $R_0 \le R\left(\vec{k}\right)/R_c \le 1$, the visco-heat modes preserve the form (Eq. (18)), they are real and completely diffusive. In this regime, there are again three cases of special interest. If $R\left(\vec{k}\right)/R_c = R_0$, then both visco-heat modes (18) are identical and equal to:

$$\lambda_{3,4} = \frac{1}{2}(\chi + \nu)k^2. \tag{21}$$

On the other hand, if the simple fluid is in a state of homogeneous thermodynamic equilibrium, $g = 0$ and $\alpha = 0$, so that $X = 0$ and $R\left(\vec{k}\right)/R_c = 0;$ consequently, in this equilibrium state (identified by the superscript e), there is a thermal diffusive mode:

$$\lambda_3^e = \chi k^2 \tag{22}$$

and the shear mode:

$$\lambda_4^e = \nu k^2. \tag{23}$$

These decay rates are well known in the literature [8, 37, 38]. Finally, because in a simple fluid, commonly ν is greater than χ, according to Eq. (18), and as $R\left(\vec{k}\right)/R_c$ grows and approaches to 1, the magnitude of the thermal diffusive mode decreases, while the shear mode grows. At the threshold of the convective regime (when $R\left(\vec{k}\right)$ reaches its critical value R_c), $R\left(\vec{k}\right)/R_c = 1$, and the two visco-heat modes (18) acquire the values:

$$\lambda_3 = 0 \tag{24}$$

and

$$\lambda_4 = (\chi + \nu)k^2. \tag{25}$$

These three cases are consistent with those obtained in analytical studies already reported for simple fluids in this regime [8, 37, 38]. Schematically, its behavior is very similar to that illustrated in **Figure 3**, and this can be seen in **Figure 1** of the reference [37].

7. Conclusions

In this work, we have used the standard formulation of *FH* to describe the dynamics of the fluctuations of a *NLC* layer in a *NESS* characterized by the simultaneous action of a uniform temperature gradient α and a constant gravitational field g, which corresponds to a Rayleigh-Bénard system. The analysis carried out takes into account only the nonconvective regime. The most important results are the analytic expressions for the seven nematic hydrodynamic modes. The explicit details of several of the calculations can be found in Refs. [25, 26]. To summarize the results obtained in this work and to put them into a proper context, the following comments may be useful.

First, in our analysis, the symmetry properties of the nematic are taken into consideration, and this allowed us to separate its hydrodynamic variables into two completely independent sets: one longitudinal, composed of five variables, and the other transverse, consisting of only two variables. From the equations that govern the dynamics of the variables in these sets, the corresponding hydrodynamic modes were calculated. The longitudinal modes are two acoustic, λ_1 and λ_2, modes (6), as well as the triplet formed by the visco-heat pair λ_3 and λ_4, modes (7), and the director λ_5, mode (8). In addition, the transverse ones are given by the shear λ_6 and the director λ_7, modes (14). We find that the influence of the temperature gradient α and the gravitational field g occurs only in the longitudinal modes, being greater its effect $(\sim 10^{-9})$ on the visco-heat pair λ_3 and λ_4. This effect is quantified by means of the Rayleigh ratio $R\left(\vec{k}\right)/R_c$, Eq. (9), where R is the Rayleigh number and R_c is its critical value above which convection sets in. The developed analysis corresponding to the nonconvective regime was carried out under the condition $R\left(\vec{k}\right)/R_c \leq 1$.

The analytical expressions calculated for the hydrodynamic modes of a nematic in the *NESS* considered exhibit behaviors that are of great interest in the following particular situations. First, if the isotropic limit of the simple fluid is taken, the *NLC* hydrodynamic modes reduce to those in the same state out of equilibrium, modes (17)–(19), [8, 37, 38]. If $R = 0$, that is, in the absence of the uniform temperature gradient and the constant gravitational field, our expressions are simplified and reduce to those already reported for a nematic in the state of thermodynamic equilibrium, modes (6), (8), (14), (15), and (16), [22, 31, 46]. In this case, if we also consider the limit of the simple fluid, they agree with those of this system in equilibrium, modes (17), (19), (22), and (23), [41, 47, 48]. When $R = R_c$, that is, at the threshold of convection, from the triplet of longitudinal λ_3, λ_4 and λ_5, the visco-heat λ_3 vanishes, and λ_4 is the sum of the thermal and shear modes, modes (12) and (13); while that of director λ_5 is identical to mode (8) [37, 38]. Moreover, if in this nematic threshold of convection, the limit of the simple fluid is considered, the modes of this system are recovered: one is zero, mode (24), and the other is the sum of the thermal and shear modes, mode (25), [37, 38]. Also, if $R\left(\vec{k}\right)/R_c < R_0\left(\tilde{k}\right)$, where $R_0\left(\vec{k}\right)$ is the reference value (10), our results predict that the visco-heat pair λ_3 and λ_4, modes (7), become propagative; in the limit of the simple fluid, under similar conditions, the corresponding modes (18) are also propagative. The latter have been predicted theoretically [8, 37, 38] and verified experimentally [43].

However, it should be mentioned that our hydrodynamic modes λ_3, λ_4, and λ_5 do not coincide with those reported in the literature for an *NLC* in the same *NESS* considered here [44, 45], which consist in one mode due to the director, another more product of the coupling of the thermal and director modes, and a shear mode. The effect of external forces α and g is only manifested in the first two modes. This triplet is reduced to the corresponding director, thermal, and shear longitudinal modes of an *NLC* in the state of thermodynamic equilibrium, as well as to the thermal and shear modes of a simple fluid in such state. It should be noted that from the analytical expressions of these modes, the existence of nematic propagative modes cannot be predicted; much less, in this *NESS*, in the simple fluid. In addition, when the threshold of convection in the nematic is considered, the director mode is canceled, another one is the sum of the thermal and director modes, and the shear mode remains unchanged; consequently, when the limit of the simple fluid is taken, they are reduced to thermal and shear modes. This last result differs completely from the already reported [37, 38] for the hydrodynamic modes of a simple fluid at the threshold of convection, where one is zero, mode (24), and the other the sum of the thermal with the shear, mode (25).

Nevertheless, our calculated expressions for the visco-heat λ_3, λ_4, and director λ_5 modes predict both the existence of propagative modes and the form that this triplet acquires in the convection threshold, and moreover, they reduce to the corresponding modes in all the different limit cases already mentioned. In this respect, we believe that they are more general than those reported in the literature [44, 45]. As far as we know, the diffusive or propagative nature of the modes λ_3 and λ_4, depending on the values taken by the ratio $R\left(\vec{k}\right)/R_c$, was not known; therefore, its derivation represents a relevant contribution of this work. Since in simple fluids, the existence of propagative modes has been predicted and verified experimentally, our predictions about the existence of this phenomenon in the modes of an *NLC* suggest the realization of new experiments.

Finally, it should be noted that this theory can be useful, since the description of some characteristics of our model lend themselves to establish a more direct contact with the experiment. Actually, physical quantities, such as director-director and density-density correlation functions, memory functions or the dynamic structure factor $S\left(\vec{k},\omega\right)$, may be calculated from our *FH* description. In Ref. [49], an application of this nature was developed by calculating the Rayleigh dynamic structure factor for the *NLC* under the *NESS* already mentioned, and its possible comparison with experimental studies is discussed; a preliminary analysis can be consulted in Ref. [50]. Another studies of the dynamic structure factor for an *NLC* in a different *NESS*, such as that produced by the presence of an external pressure gradient, were published in the references [19, 20].

Acknowledgements

We thank the UACM for the economic facilities granted to cover the total cost of the publication of this research work.

Conflict of interest

The authors declare that they have no conflict of interests.

Nomenclature

k	wave number
ω	angular frequency
τ	relaxation time of almost all degrees of freedom
$S\left(\vec{k},\omega\right)$	dynamic structure factor
$\vec{k}$	wave vector
R	Rayleigh number
R_c	Rayleigh number at the convection threshold
R/R_c	Rayleigh ratio
$O(3)$	orientation symmetry group
$T(3)$	translation symmetry group
$\hat{n}$, $\vec{n}$ or n_α	director field
d	thickness of the nematic layer

$\vec{g}$	constant gravitational force of magnitude g
$\hat{x}, \hat{y}, \hat{z}$	Cartesian unitary vectors
x, y, z	Cartesian coordinates
T	temperature
α	temperature gradient of magnitude $\nabla_z T$
p	hydrostatic pressure
$\nabla_z p$	pressure gradient
ρ	volumetric density of mass
$\vec{v}$	flow velocity
s	specific density of entropy (entropy per unit mass)
$\vec{r}$	position vector
ΔT	temperature difference between the plates of the cell
X	effective temperature gradient
β	coefficient of thermal expansivity
c_p	specific heat at constant pressure
c_v	specific heat at constant volume
γ	ratio of specific heats
c_s	adiabatic sound velocity
c_T	isothermic sound velocity
Ψ	set of nematodynamic variables
$\delta\varphi$	divergence of $\delta \vec{v}$
$\delta\psi$	component z of the rotational of $\delta \vec{v}$
$\delta\xi$	component z of the double rotational of $\delta \vec{v}$
δf_1	divergence of $\delta \vec{n}$
δf_2	component z of the rotational of $\delta \vec{n}$
t	as superscript, indicates transpose matrix
$\delta \vec{X}\left(\vec{k},t\right)$	vector whose components are the spatial Fourier transform of the variables δp, $\delta\varphi$, δs, $\delta\psi$, $\delta\xi$, δf_1 and δf_2
$\delta \vec{X}^L\left(\vec{k},t\right)$	longitudinal component of $\delta \vec{X}\left(\vec{k},t\right)$
$\delta \vec{X}^T\left(\vec{k},t\right)$	transverse component of $\delta \vec{X}\left(\vec{k},t\right)$
M	coefficient matrix of the linear system for $\delta \vec{X}\left(\vec{k},t\right)$
M^L and M^T	longitudinal and transverse submatrices of M
$\vec{\Theta}\left(\vec{k},t\right)$	stochastic vector of the linear system for $\delta \vec{X}\left(\vec{k},t\right)$
$\vec{\Theta}^L\left(\vec{k},t\right)$	longitudinal component of $\vec{\Theta}\left(\vec{k},t\right)$
$\vec{\Theta}^T\left(\vec{k},t\right)$	transverse component of $\vec{\Theta}\left(\vec{k},t\right)$
$z_i\left(\vec{k},t\right)$	variables of same dimensionality ($i = 1,\ldots, 7$)
$\vec{Z}\left(\vec{k},t\right)$	vector of the variables $z_i\left(\vec{k},t\right)$
$\vec{Z}^L\left(\vec{k},t\right)$	longitudinal component of $\vec{Z}\left(\vec{k},t\right)$
$\vec{Z}^T\left(\vec{k},t\right)$	transverse component of $\vec{Z}\left(\vec{k},t\right)$
N	coefficient matrix of the linear system for $\delta \vec{Z}\left(\vec{k},t\right)$
N^L and N^T	longitudinal and transverse submatrices of N
$\vec{\Xi}\left(\vec{k},t\right)$	noise vector of the linear system for $\vec{Z}\left(\vec{k},t\right)$

$\vec{\Xi}^L\left(\vec{k},t\right)$	longitudinal component of $\vec{\Xi}\left(\vec{k},t\right)$
$\vec{\Xi}^T\left(\vec{k},t\right)$	transverse component of $\vec{\Xi}\left(\vec{k},t\right)$
ζ_i	noise components of $\vec{\Xi}^L$ $(i = 1,\dots, 5)$ and $\vec{\Xi}^T$ $(i = 6, 7)$
$p(\lambda)$	characteristic polynomial of the matrix N
$p^L(\lambda)$	characteristic polynomial of the submatrix N^L
$p^T(\lambda)$	characteristic polynomial of the submatrix N^T
λ	eigenvalues of $p(\lambda)$
$\vec{Z}_X^L\left(\vec{k},t\right)$ and $\vec{Z}_Y^L\left(\vec{k},t\right)$	components of the vector $\vec{Z}^L\left(\vec{k},t\right)$
$p_{XX}^L(\lambda)$ and $p_{YY}^L(\lambda)$	polynomials in which $p^L(\lambda)$ is broken down
D_T	anisotropic thermal coefficient
σ_1, σ_2, σ_3, and σ_4	anisotropic viscous coefficients
$\chi_\parallel$ and $\chi_\perp$	thermal diffusivities parallel and perpendicular to $\vec{n}$
χ_a	anisotropic thermal diffusivity
ν_i	nematic viscosities $(i = 1, \dots, 5)$
Ω and λ_+	anisotropic adimensional nematic coefficients
K_1, K_2 and K_3	elastic coefficients of Frank
K_I and K_{II}	anisotropic elastic coefficients
γ_1	torsion viscosity
ϖ	auxiliary parameter
a_0, a_0', and a_0''	small dimensionless longitudinal quantities
a_1, a_1', and a_2	small dimensionless longitudinal quantities
a_3, a_5, and a_6	small dimensionless longitudinal quantities
λ_1 and λ_2	acoustic propagative longitudinal modes
Γ	anisotropic sound attenuation coefficient
λ_3 and λ_4	visco-heat longitudinal modes
λ_5	director diffusive longitudinal mode
$k_\perp$and $k_\parallel$	components of $\vec{k}$ perpendicular and parallel to $\hat{n}_0$
$\hat{k}_\perp \equiv k_\perp/k$	unit vector of $\hat{k}_\perp$
R_0	reference value of the Rayleigh ratio below which visco-caloric modes are propagative
a_4, a_5', and a_6'	small dimensionless transverse quantities
λ_6 and λ_7	shear and director diffusive transverse modes
λ_i^e	nematic modes in the state of equilibrium $(i = 1, \dots, 7)$
χ	thermal diffusivity of a simple fluid
η and ζ	shear and volumetric viscosities of a simple fluid
ν	kinetic viscosity of a simple fluid
Γ'	attenuation coefficient of sound in a simple fluid

Author details

Jorge Fernando Camacho[1,3*] and Rosalío Fernando Rodríguez[2,3,4]

1 Universidad Autónoma de la Ciudad de México (UACM), Mexico City, Mexico

2 Instituto de Física, Universidad Nacional Autónoma de México (UNAM), Mexico City, Mexico

3 Sistema Nacional de Investigadores (SNI), México

4 Proyecto Universitario de Fenómenos Nolineales y Mecánica (FENOMEC), UNAM, México

*Address all correspondence to: fernando.camacho@uacm.edu.mx

References

[1] Landau LD, Lifshitz E. Fluid Dynamics. New York: Pergamon; 1959

[2] Landau LD, Lifshitz E. Hydrodynamic fluctuations. Journal of Experimental and Theoretical Physics. 1957;**32**:618-619

[3] Fox RF, Uhlenbeck GE. Contributions to non-equilibrium thermodynamics. I. Theory of hydrodynamical fluctuations. Physics of Fluids. 1970;**13**:1893-1902

[4] Fox RF, Uhlenbeck GE. Contributions to non-equilibrium thermodynamics. I. Theory of hydrodynamical fluctuations. Physics of Fluids. 1970;**13**:2881-2890

[5] Fox RF. Gaussian stochastic processes in physics. Physics Reports. 1978;**48**:179-283

[6] Onsager L, Machlup S. Fluctuations and irreversible processes. Physics Review. 1953;**91**:1505-1512

[7] Onsager L, Machlup S. Fluctuations and irreversible processes. Physics Review. 1953;**91**:1512-1518

[8] Ortíz de Zárate JM, Sengers JV. Hydrodynamic Fluctuations in Fluids and Fluid Mixtures. Amsterdam: Elsevier; 2006. 309 p

[9] Dorfman JR, Kirkpatrick TR, Sengers JV. Generic long-range correlations in molecular fluids. Annual Review of Physical Chemistry. 1994;**45**:213-239

[10] Grinstein G, Lee DH, Sachdev S. Conservation laws, anisotropy, and "self-organized criticality" in noisy nonequilibrium systems. Physical Review Letters. 1990;**64**:1927-1930

[11] Ronis D, Procaccia I, Machta J. Statistical mechanics of stationary states. VI. Hydrodynamic fluctuation theory far from equilibrium. Physical Review A. 1980;**22**:714-724

[12] Beysens D, Garrabos Y, Zalczer G. Experimental evidence for Brillouin asymmetry induced by a temperature gradient. Physical Review Letters. 1980; **45**:403-406

[13] Law BM, Sengers JV. Fluctuations in fluids out of thermal equilibrium. Journal of Statistical Physics. 1989;**57**: 531-547

[14] Li WP, Segrè PN, Gammon RW, Sengers JV. Small-angle Rayleigh scattering from nonequilibrium fluctuations in liquids and liquid mixtures. Physica A: Statistical Mechanics and its Applications. 1994; **204**:399-436

[15] Fuller G, van Egmond J, Wirz D, Peuvrel-Disdier E, Wheeler E, Takahashi H. Enhancement of concentration fluctuations in solutions subject to external fields. In: Nakatani I, Dadmum MD, editors. Flow-Induced Structure in Polymers ACS Symposium Series. Vol. 597. Washington DC: American Chemical Society; 1995. pp. 22-34

[16] Sengers JV, Gammon RW, Ortíz de Zárate JM. Thermal-diffusion driven concentration fluctuations in a polymer solution. In: Dadum M, Noid D, Sumpter B, Melnichenko Y, editors. Computational Studies, Nanotechnology and Solution Thermodynamics. New York: Kluwer; 2000. pp. 37-44

[17] Pleiner H, Brand H. Light scattering in nematic liquid crystals in a nonequilibrium steady state. Physical Review A. 1983;**27**:1177-1183

[18] Pleiner H, Brand H. Light scattering in nematic liquid crystals in the

presence of shear flow. Journal de Physique Lettres. 1983;**44**:23-31

[19] Rodríguez RF, Camacho JF. Non-equilibrium effects on the light scattering spectrum of a nematic driven by a pressure gradient. Revista Mexicana de Fisica. 2002;**48**(S1): 144-153

[20] Camacho JF, Híjar H, Rodríguez RF. Hydrodynamic correlation functions for a nematic liquid crystal in a stationary state. Physica A: Statistical Mechanics and its Applications. 2005;**348**:252-276

[21] Kadanoff LP, Martin PC. Hydrodynamic equations and correlation functions. Annals of Physics. 1963;**24**:419-469

[22] Pleiner H, Brand HR. Hydrodynamics and electrohydrodynamics of liquid crystals. In: Buka A, Kramer L, editors. Pattern Formation in Liquid Crystals. Berlin: Springer-Verlag; 1996. pp. 15-67

[23] Stephen MJ. Hydrodynamics of liquid crystals. Physical Review A. 1970; **2**:1558-1562

[24] Camacho JF. Hidrodinámica fluctuante de cristales líquidos nemáticos [PhD: Thesis]. Mexico City: Universidad Nacional Autónoma de México; 2017

[25] Camacho JF, Rodríguez RF. Hydrodynamic coupled modes of a nematic under a temperature gradient and a uniform gravitational field. arXiv: 1604.0825v1 [cond-mat. soft]. 2016:1-29

[26] Camacho JF, Rodríguez RF. Nonequilibrium fluctuations from a nematic under a thermal gradient and a gravity field. I. Hydrodynamic modes. The European Physical Journal B. 2017; **90-21**:1-11

[27] Zwanzig R. Time-correlation functions and transport coefficients in statistical mechanics. Annual Review of Physiology. 1965;**16**:67-102

[28] DA MQ. Statistical Mechanics. New York: Harper and Row; 1976. 641 p

[29] Brillouin L. Diffusion de la lumière et des rayons X par un corps transparent homogène. Annales de Physique. 1922; **9**(17):88-122

[30] Brillouin L. Über die Fortpflanzung des Lichtes in dispergierenden Medien. Ann. Phys. 1914;**349**:203-240

[31] Kats EI, Lebedev VV. Fluctuational Effects in the Dynamics of Liquid Crystals. Berlin: Springer; 1994. 170p

[32] de Gennes PG, Prost J. The Physics of Liquid Crystals. Oxford: Clarendon Press; 1995. 597 p

[33] Forster D. Hydrodynamic Fluctuations, Broken Symmetry and Correlation Functions. Reading MA: W. A. Benjamin Inc.; 1975. 343 p

[34] Groupe d'etude des cristaux liquides (Orsay). Dynamics of fluctuations in nematic liquid crystals. The Journal of Chemical Physics. 1969;**51**:816

[35] Chandrasekhar S. Liquid Crystals. 2nd ed. New York: Cambridge; 1992

[36] Reyes JA, Rodríguez RF. Propagation of optical solitons and fields in nonlinear nematic waveguides. In: Chen LV, editor. Focus on Soliton Research. New York: Nova Science Publishers; 2006. pp. 79-136

[37] Segrè PN, Schmitz R, Sengers JV. Fluctuations in inhomogeneous and nonequilibrium fluids under the influence of gravity. Physica A: Statistical Mechanics and its Applications. 1993;**195**:31-52

[38] Lekerkerker HNW, Boon JP. Hydrodynamic modes and light scattering near the convective

instability. Physical Review A. 1974;**10**: 1355-1360

[39] Chandrasekhar S. Hydrodynamic and Hydromagnetic Stability. New York: Dover; 2003. 704 p

[40] Mountain RD. Spectral distribution of scattered light in a simple fluid. Reviews of Modern Physics. 1966;**38**: 205-214

[41] Berne BJ, Pecora R. Dynamic Light Scattering. New York: Wiley; 1976. 376 p

[42] Pieranski P, Dubois–Violette E, Guyon E. Heat convection in liquid crystals heated from above. Physical Review Letters. 1973;**30**:736-739

[43] Boon JP, Allain C, Lallemand P. Propagating thermal modes in a fluid under thermal constraint. Physical Review Letters. 1979;**43**:199-203

[44] Miyakawa K. Fluctuations in nematic liquid crystals under the thermal gradient. Journal of the Physical Society of Japan. 1975;**39**:628-633

[45] Migranov NG, Verevochnikov AV, Chuvyrov AN. Hydrodynamic fluctuations of the main variables in nematic liquid crystals subjected to the temperature gradient. Molecular Crystals and Liquid Crystals. 1998;**319**: 31-37

[46] Leslie FM. Some constitutive equations for anisotropic fluids. The Quarterly Journal of Mechanics and Applied Mathematics. 1966;**19**:357-370

[47] Boon JP, Yip S. Molecular Hydrodynamics. New York: Dover; 1991. 417 p

[48] Benedek GB. Thermal fluctuations and the scattering of light. In: Chrétien M, Deser S, Groos EP, editors. Statistical Physics, Phase Transitions and Superfluity, Brandeis University Summer Lectures. Vol. 2. New York: Gordon and Breach; 1966. p. 1

[49] Camacho JF, Rodríguez RF. Nonequilibrium fluctuations from a nematic under a thermal gradient and a gravity field. II. Rayleigh scattering. The European Physical Journal B. 2017; **90-22**:1-11

[50] Rodríguez RF, Camacho JF. Nonequilibrium thermal light scattering from nematic liquid crystals. In: Macias A, Díaz E, Uribe F, editors. Recent Developments in Mathematical and Experimental Physics, Vol. B Statistical Physics and Beyond. New York: Kluwer; 2003. pp. 197-214

Chapter 7

Non-Newtonian Dynamics with Heat Transport in Complex Systems

Aamir Shahzad and Fang Yang

Abstract

Transport properties of complex system under various conditions are of practical interest in the field of science and technology. Homogenous nonequilibrium molecular dynamics (HNEMD) simulations have been employed to calculate the thermal conductivity (λ) of three-dimensional (3D) strongly coupled complex nonideal plasmas (SCCNPs) over a suitable range of plasma parameters (Γ, κ). New investigations show that the λ depending on plasma parameters and minimum value of λ exists at nearly same plasma states. In the present case, the non-Newtonian behavior is checked with different system sizes and it is found that the λ behavior is well matched with earlier numerical work. It is demonstrated that the present outcomes are more consistent than those obtained earlier known simulations. It is revealed that our outcomes can be acceptable for a low range of force field in order to find out the size of linear ranges, and it explains the nature of nonlinearity of SCCNPs. It has been shown that the measured outcomes at steady states of external field of F^* (=0.005) are in acceptable agreement with previous numerical outcomes, and it showed that the deviations are within less than 12% for most of the data and depend on plasma states.

Keywords: non-Newtonian, thermal conductivity, homogenous nonequilibrium molecular dynamics, strongly coupled complex nonideal plasmas, external force field

1. Introduction

The computational knowledge of thermophysical properties is very different of complex liquids as compared to the nonionic liquids. The important thermal conductivity of complex liquids is used in the heat design process as an important parameter. The estimations of thermal conductivity obtained by applying the molecular dynamics (MD) approach in liquids and crystalline solids are a difficult job due to perceived limitations of computational power [1]. Recently, in the field of science and technology, the transport properties of interacting particles in complex nonideal systems are of practical importance. A deep understanding of the interaction of complex systems is required for nano- and microstructuring of surfaces. The micron-size particles have recently been investigated in complex (dusty) plasmas, and in the physics and chemistry of plasmas, space environment, ionized gases, and material research and in the nuclear energy generation. The complex (dusty) plasmas play very important role in various technological applications, such as industrial processing of microelectronic devices, storage devices, and fuel burning, and future energy production [2]. For the explanation and understanding of these macroscopic phenomena, a comprehensive microscopic knowledge and calculation of the transport properties of complex (dusty) plasmas are required over the extensive range of plasma parameters (Γ, κ). Both the Coulomb coupling (Γ) and Debye screening strength (κ) are the dimensionless parameters, which can be used to characterize the plasma. In statistical mechanics, the microscopic dynamical origin of heat transport is a fundamental problem. Moreover, the purpose of the present work is to investigate the thermal conductivity dependences on the strength of different perturbation fields and to understand the non-Newtonian behaviors in the Yukawa liquids along with thc calculations of thermal conductivity.

1.1 Dusty plasmas

Dusty plasmas are also known as nonideal complex plasmas that contain particulates of condensed matter. The dust particles may have sizes ranging from nanometers to micrometers, and typically much more massive than that of plasma ions, electrons, and neutrons. When the dust particles immersed in the plasma, they attain a high electric charge (negative charge) which makes the dusty plasmas interesting and technological important in the area of applied plasma physics. The dynamical behavior of these massive dust charge particles is much complex and occurs on considerably slower time scales, because their charge-to-mass ratio is in orders of magnitude smaller than that of the corresponding charge-to-mass ratio of either the electrons or ions. Dust particles are found in the large abundant in planetary plasmas, cosmic plasmas, plasmas in the laboratory, and plasmas near the earth. As a matter of fact, one may cogitate that except in the hottest regions of fusion plasmas, where the dust particles would not survive, most are known as dusty plasmas in the sense that some dust particles might be present. To understand this fact, one recognizes two cases in which: (1) there are just few secluded (noninteracting) dust particles, with the goal that they have nearly nothing if any impact on the plasma, and (2) there are countless number of dust particles in the plasma so that their existence really changes the properties and behavior of the plasma. In the event (1), the dust particles are charged by their interactions with the plasma yet do not change the plasma in any noticeable way. Then, again case (2) agrees to the situation in which the charge dust is a component of the plasma, subject to the collective interactions that recognize an ionized gas from a neutral gas. Case (2) is what is ordinary characterized as "dusty plasmas." At a significantly bigger scale, it is outstanding that comets for the most part have two tails. One tail is expected to the comet's dust particles, the other is because of ionized gas comet coma. These tails are not separate near the coma but overlap forming dusty plasma [3].

1.2 Dusty plasma parameters

There are two basic dimensionless parameters which are used for the analysis of transport coefficients in 2D and 3D dusty plasma systems, and which are responsible for mass transfer and phase state in nonideal dissipative systems [4–6]. These dimensionless parameters are known as effective Coulomb coupling (Γ^*) parameter and scaling parameter (ξ). These parameters (Γ^*, ξ) are responsible for transport and structural processes for nonideal systems [5]. Screening parameter (κ) is the third one which is the important for the classification of dusty plasmas. Here, we discuss only two parameters, Coulomb coupling (Γ) and the Debye screening strength (κ).

1.2.1 Coulomb coupling parameter (Γ)

The Coulomb coupling (Γ) parameter is the ratio between the interparticle potential energy (P.E) to kinetic energy (K.E), and mathematically, it is written as, $\Gamma = \frac{P.E}{K.E}$. This Coulomb coupling (Γ) parameter is used for the classification of strongly coupled complex (dusty) plasmas and weakly coupled complex (dusty) plasmas.

1.2.2 Screening parameter (κ)

Another important parameter of dusty plasmas is the screening strength (κ), which is the ratio of interparticle distance to the Debye length, and mathematically, it is written as $\kappa = \frac{a}{\lambda_D}$, here "$a$" is the Wigner-Seitz (WS) radius and "λ_D" is the Debye length. The screening parameter (κ) is also used for the classification of dusty plasmas.

1.3 Types of dusty plasmas

On the basis of Coulomb coupling (Γ) parameter, the complex (dusty) plasmas are classified into two classes: one is called weakly coupled (ideal) complex (dusty) plasmas and the other is called strongly coupled (nonideal) complex (dusty) plasmas. These are succinctly discussed below.

1.3.1 Weakly coupled complex (dusty) plasmas

For weakly coupled complex (dusty) plasmas ($WCCDP_S$), the Coulomb coupling parameter (Γ) is less than the unity ($\Gamma < 1$), and it also called the weakly coupled ideal plasmas because Columbic collisions are negligible. Weakly coupled plasmas, like a gas, have no structure because Coulombic interactions are negligible between the particles and particle motion is like molecular motion in gases and particles have nearly random positions with respect to nearest neighbors [7]. For $WCCDP_S$, the K.E (thermal energy) is much larger than the Coulomb interparticle P.E of nearest particles.

1.3.2 Strongly coupled complex (dusty) plasmas

When the Coulomb coupling parameter (Γ) is greater than the unity ($\Gamma > 1$), then the plasma is known as strongly coupled (nonideal) complex plasmas. For weak-to-intermediate Coulomb coupling (Γ) values, the $SCCDP_S$ can have structure of liquids, and structure of solids for higher values of Γ. Furthermore, $SCCDP_S$ are the collection of free microparticles that interact with each other with a strong Coulomb repulsion force and have structure at microscopic scale for the arrangement of particles. The particle motion in $SCCDP_S$ resembles that in the liquids or solids, and particles remain in relative fixed positions with respect to neighboring particles because of the strong Coulomb interactions present between the charged particles [7]. In $SCCDP_S$, the Coulomb interaction P.E is much larger than the K.E of nearby particles.

1.4 Formation and growth of dust particles in a plasma

An innovative feature of plasmas is that comprise chemically energetic species which grow the dust particles. It is true for plasma appliances that are using in the plasma processing semiconductor engineering, in which combinations of gases such as oxygen (O_2), argon (Ar), and silane (SiH_4) are castoff in the assembly and figments of microelectronic chips. Dust particles could also produce through sputtering, arcing, electron bombardment, etc., where the atoms or molecules from the walls of chamber or electrodes come out and immersed in the plasmatic system through a different mechanism, called plasma-material interaction [8]. Plasma processing devices are employed for the production of silicon wafers characteristically used as parallel plate electrode, in which 13.56 MHz radio frequency (rf) power is connected to the lower electrode to create the plasma. Etching process involves a reactive species such as silane along with a buffer gas like Ar. Plasma-aided gas phase chemical reactions produced silicon hydride (SiH_2) by the reaction $\{e^- + SiH_4 \rightarrow (SiH_4)^* \rightarrow SiH_2 + 2H\}$. The vibrationally excited state is produced by the collisions of SiH_4 with electrons which then dissociates into SiH_2 [9]. The particle that grows in plasmas passes through certain phases like nucleation, coagulation, and surface growth.

1.5 Dust particle in the plasma

Dust particles are found everywhere in the entire universe with different shapes and sizes and mostly found in the solid form but also found in the liquid and gaseous ionized form. When the dust particles coexist with the plasma, then "dusty (complex) plasma" is formed. Dust particles acquire an electric negative charge, when these dust particles are immersed in the plasma and then affected by electric and magnetic fields and plasma properties are changed. Moreover, these dust particles attain electric negative charge (typically depends on the flow of ions and electrons) very fast due to the interactions between the dust particles and the nearby plasmas.

In recent years, dusty plasmas have opened up an entirely new field of research of science and technology by investigating transport properties (thermal conductivity, shear viscosity, and diffusion coefficient) of dusty plasmas in the laboratory.

1.5.1 Charge on dust particle

A lot of mechanisms are adopted to produce charge on a dust particle. If all mechanisms are considered at once, then the measurement of the equilibrium charge condition on a grain becomes very difficult. For the electron temperature T_e and ions temperature T_i, the flux of ions and electrons has an individual thermal velocity v_{te}. The thermal velocity of electron is higher than that of the heavier ions which have a minute thermal velocity v_{ti}. These motions develop the charge Q on the grain and make its surface potential (ϕ_s) negative. The charge on a grain fluctuates continuously due to the collective current of electron and ion, i.e., $dQ/dt = I_e + I_i$ and equilibrium of charge occurs under condition $I_e + I_i = 0$ or $\phi_s = -2.49kT/e$ for an electron-ion plasma. The charge itself is associated to the surface potential by $Q = C\phi_s$, in which C tells about the capacitance of a grain in a plasma [10]. The dust particle density differs from the density of electrons and ions because in normal plasma, neutral n_0 exists. If the primary electrons are very energetic, then they release subordinate electrons, which make the positive potential surface. The absorption of plasma ions also makes the positive potential surface. The transition of primary electrons and ions influenced by the surface potential of the grain depends on the relative velocity between the plasma species and the grain. Electrons are repelled and the grain current will be decreased if surface potential is negative. Electrons show attraction and the grain current increases for positive potential surface.

1.5.2 Size of dust particle

In dusty plasma, the dust particles can have any shape and can be made of either dielectric or conducting materials. The size of dust particles is much larger than the size of electrons and ions of plasmas which is in microns or tens of nanometers. So, the dust particles can easily be seen without any microscope. The typical size range of dust particles is from 100 nm up to about 100 μm. For experimental studies, dust particles that are distributed into plasmas are generally plastic or glass particles (commonly used particle is melamine formaldehyde). They are spherical in shape with a very narrow distribution of diameters. For example, the diameter of a normally used particle may be 3.50 ± 0.05 μm and a mass $\sim 3 \times 10^{-11}$ kg. Such particles are named monodisperse. Fine powders, such as aluminum silicate (kaolin) with a broad size distribution ranging from the submicron to tens of microns, are used in some experiments. Such powders contain particles having laminar shapes with jagged edges.

1.6 Forces acting on dust particles in a plasma

When the dust particles are immersed in the plasma, then various forces are acting on the dust particles which are significant because of their dynamics and transport characteristics. These various forces determine that where the particles are trapped or not, and these are sensitive to the position of dust particles within the plasmas.

1.6.1 Force of gravity ($\boldsymbol{F}_g$)

The dust particles which are under a gravitational force are proportional to the mass of dust particle, and if dust particles are under microgravity condition, then it must be ignored.

$$F_g = mg_d = \frac{4}{3}\pi a^3 \rho_d g \tag{1}$$

where "g" describes the gravitational acceleration and "ρ_d" represents the mass density of dust particles. Mostly, its value for solid materials is ρ_d = 2000 kg/m^3. This force can be neglected for submicron particles, but for the micron sized or larger particles, this can be considered as the dominant force that typically confines the time during which the particle resides in the plasmas.

1.6.2 The electric force ($\boldsymbol{F}_e$)

For plasma having an electric field (E), the electric force acting on dust species is written as:

$$F_e = QE \tag{2}$$

This force is much larger in the bulk of the plasmas, while it is much smaller in the sheaths next to the plasma edge. For example, for characteristic radio frequency (rf) parallel plate glow discharge plasmas, particles reside beneath the negative electrode where the downward electric field offers an upward electric force that stabilizes the weight of the particle with force.

1.6.3 Neutral drag force (F_n)

This force is generated due to impacts of dust particles with the neutral gas species (atoms and molecules) and it is proportional to the neutral pressure in the vacuum chamber. Mathematically, it is written as:

$$F_n = N m_n {v^2}_{dn} \pi a^2 \quad (3)$$

where N defines the density of neutral species, m_n denotes the mass of neutral species, and v_{dn} represents the average relative velocity between the neutral elements and dust species. The resulting damping force also acts on the dust particles if the dust particles drift with drag force in the opposite direction to its motion.

1.6.4 Thermophoretic force (F_{th})

The thermophoretic force occurs from the effect of temperature gradient in the neutral gas in the plasmas, and it is in the opposite direction to the temperature gradient. This force occurs due to transfer of momentum by the gas molecules from the hotter region to the colder portion of the gas. It can be written as:

$$F_{th} = \frac{16\sqrt{\pi}}{15} \frac{a^2 \kappa_T}{\nu_{T,n}} \nabla T_n \quad (4)$$

where $v_{T,n}$ describes the thermal speed of the neutral gas of plasma, κ_T defines the translational effects in the λ, and T_n tells about the temperature of the neutrals. It can be occurred in the discharge by heating one of the electrodes. The force of gravity acting on the dust particles in plasma is balanced by the temperature gradient [11].

1.7 Application of dusty plasmas in industry

Plasma-based material processing technologies are extensively employed in the designing and commercialization of very large-scale integrated circuits (VLSI). Usually, chemically reactive plasmas are useful to sputter, etch, or otherwise alter the surface characteristics especially for silicon. Surface characteristics are approximately done at length scale of 0.2 μm wide and 4 μm deep in silicon films by such kind of mechanisms. The presence of dust is of critical concern to the microelectronic industry since particle contamination of semiconductor materials was estimated to account for more than 50% of device failures. Dust contamination diminishes the yield and recital characteristics of fabricated devices. Simply, the dust particles fall into the surface topographies of semiconductors either interfering with the etching process, preventing the adhesion of thin films or contaminating the final products. The occurrence of even the smallest dust particles became a crucial problem as the microelectronics industry moved to smaller and smaller structures.

1.7.1 Dust is a good thing

In those days, it was investigated that dust particles in plasma can have very interesting and useful properties, e.g., very small sizes, uniform size distribution, and chemical activity. There are many applications of plasma-produced particles. For example, large and active surface in catalysis is profitable. They are also essential in ceramic industry for sintering, in the modern technology of composite materials, and in fabrication of hard coatings [12] and solar cells [13]. Also, by injecting particles in plasma can furnish unique objects, like coated or layered grains with desired surface structure, color, and fluorescent properties. These particles are used as toners in copying machines [14] or in some optical devices [15].

1.7.2 Dust in plasma processing devices (dust is a bad thing)

During the last decade, in microelectronics industry, dust particles become the major cause of contamination and reduce the yield and performance of fabricated devices. In early 1990s, more than 50% devices were failed due to the particle contamination. The adhesion of thin films was reduced due to the submicron particles deposited on the surface, also causes dislocations. In semiconductor technology, the elimination of even smallest dust particles has become an urgent issue to develop smaller structures and thin films. Firstly, it was thought that the cause of most of the contamination is that the processed surfaces were handling in the clean rooms but soon it was seemed that plasma is the major source of dust particles and causes the loss of costly wafers. Now, the dust contamination is well controlled.

2. HNEMD algorithm and computational technique

We start, as usual, the Green-Kubo relations (GKR$_S$) for the hydrodynamic transport coefficients of uncharged particles [16]. It is well-known form and has been shown the standard GKR$_S$ of fluids to the YDPL$_S$ [17–22]. The typical GKR$_S$ used for the estimation of thermal conductivity of interacting dust particles for YDPL$_S$:

$$\lambda = \frac{1}{3k_B V T^2} \int_0^\infty \langle \mathbf{J}_Q(t).\mathbf{J}_Q(0) \rangle dt \tag{5}$$

where in Eq. (5), k_B is the Boltzmann's constant, V is the system volume, T is the absolute temperature, and $\mathbf{J_Q}$ is the heat flux vector. The expression for the microscopic heat flux vector $\mathbf{J_Q}$ can be given by:

$$\mathbf{J}_Q V = \sum_{i=1}^{N} E_i \frac{\mathbf{p}_i}{m} - \frac{1}{2} \sum_{i \neq j} (\mathbf{r}_i - \mathbf{r}_j) \left(\frac{\mathbf{p}_i}{m} . \mathbf{F}_{ij} \right) \tag{6}$$

In the above expression, $\mathbf{F}_{ij}$ is the total interparticle force on particle i due to j, $\mathbf{r}_{ij} = \mathbf{r}_i - \mathbf{r}_j$ are the position vectors, and $\mathbf{P}_i$ is the momentum vector of the ith particle. E_i is the energy of particle i and is given by the expression as:

$$E_i = \frac{\mathbf{p}^2{}_i}{2m} + \frac{1}{2} \sum_{i \neq j} \phi_{ij} \tag{7}$$

where ϕ_{ij} is the Yukawa pair potential between particle i and j. Evans [23] has developed the non-Hamiltonian linear response theory (LRT) used for a moving system representing the equation of motion:

$$\dot{\mathbf{r}}_i = \frac{\mathbf{p}_i}{m} \tag{8}$$

$$\dot{\mathbf{p}}_i = \sum_{J=1}^{N} \mathbf{F}_i + \mathbf{D}_i(\mathbf{r}_i, \mathbf{p}_i).\mathbf{F}_e(t) - \alpha \mathbf{p}_i \tag{9}$$

where $\mathbf{F}_i = \left(-\partial\phi_{ij}/\partial r_i\right)$ is the total interparticle force acting on particle i and $\mathbf{D}_i = \mathbf{D}_i(\mathbf{r}_i, \mathbf{p}_i)$ is the tensor phase variable that describes the coupling of system to the fictitious external force field $\mathbf{F}_e(t)$. A mechanical work is performed through the external applied force field $\mathbf{F}_e(t)$, and thus, the equilibrium cannot be maintained. In the above Eq. (9), α is the Gaussian thermostat multiplier that maintains the system temperature and it is given as [16, 23]:

$$\alpha = \frac{\sum_{i=1}^{N}\left[\mathbf{F}_i + \mathbf{D}_i(\mathbf{r}_i, \mathbf{p}_i).\mathbf{F}_e(t)\right].\mathbf{p}_i}{\sum_{i=1}^{N} p_i^2/m_i} \tag{10}$$

The external force field parallel to the z-axis is of the form $\mathbf{F}_e(t) = (0, 0, F_Z)$; therefore, the thermal conductivity is calculated as:

$$\begin{aligned} \lambda &= \frac{V}{3k_B T^2} \int_0^{\infty} \langle \mathbf{J}_{Q\,z}(t)\ \mathbf{J}_{Q\,z}(0) \rangle dt \\ &= \lim_{F_z \to 0} \lim_{t \to \infty} \frac{-\left\langle \mathbf{J}_{Q_z}(t) \right\rangle}{T F_z} \end{aligned} \tag{11}$$

where $\mathbf{J}_{Q_Z}(t)$ is the z-component of the heat flux vector and the external force field $\mathbf{F}_e(t) = (F_Z)$.

3. Numerical results and discussion

This section provides the outcomes of thermal conductivity of 3D complex dusty plasmas by using HNEMD simulations over suitable plasma couplings Γ ($\equiv$1, 200) and screening strengths κ ($\equiv$1.4, 4) at constant external force strength of F_{ext} ($\equiv$0.005). It is noted that we have already reported our similar results with higher system sizes [13] and with different varying force fields [13]. In this present work, we have reported our HNEMD outcomes for different low to intermediate system sizes at fixed force field.

Figures 1–4 show our main outcomes of plasma thermal conductivity (λ) by employing HNEMD approach. Here, the thermal conductivity is normalized by plasma frequency (ω_p) as $\lambda_0 = \lambda/nk_B\omega_p a_{ws}$, or by the Einstein frequency (ω_E) as $\lambda^* = \lambda/\sqrt{3}nk_B\omega_E a_{ws}$ of SCCNPs, at the normalized external field strength $F^* = (F_z)(a_{ws}/J_Q)$, where a_{ws} is radius of Wigner-Seitz (WS) radius with n being the equilibrium particle number density, k_B is Boltzmann constant and J_Q. It should be noted that these normalizations have been employed for classical Coulomb one-component plasmas (COCPs) [24] and SCCNPs [7].

Diverse series of the plasma λ_0 subsequent to a decreasing series of external force field F^* are computed to establish the linear system of the SCCNPs under the influence of the normalized force field strength. The current HNEMD outcomes allow investigation for the complete range of plasma parameters (Γ, κ) with variation of external force field F^*. In this work, a feasible suitable value of the external

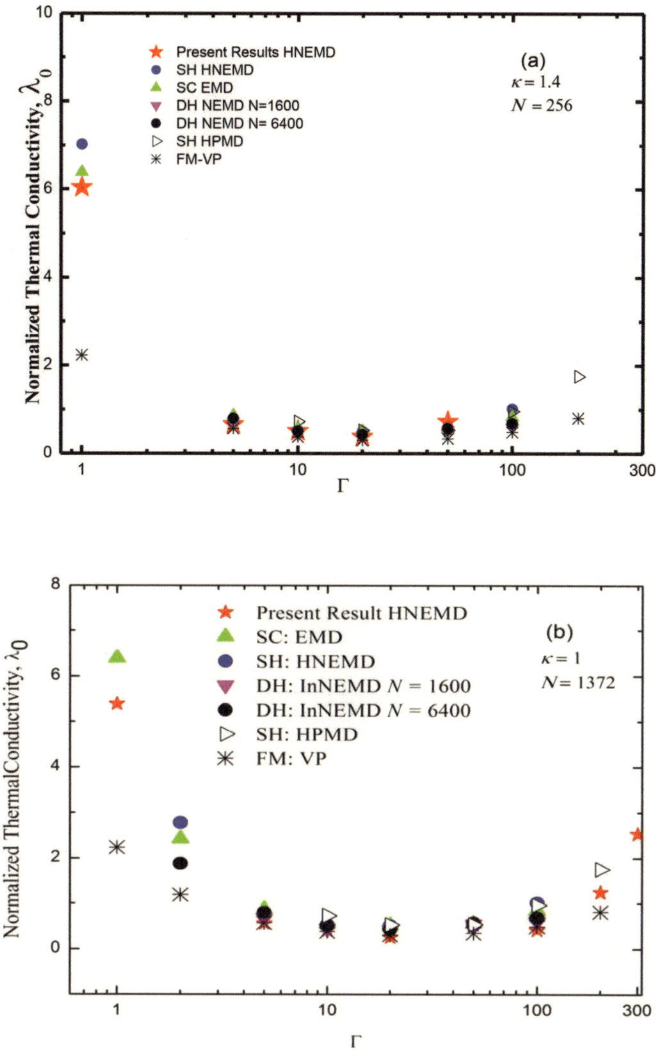

Figure 1.
Comparison of normalized plasma thermal conductivity (λ_o), computed by various numerical approaches for plasma coupling states $\Gamma \equiv (1, 200)$. Results investigated by Shahzad and He for homogenous nonequilibrium MD (HNEMD) [13], Salin and Caillol for equilibrium MD (EMD) [21], Donko and Hartmann for inhomogenous NEMD [17]. Shahzad and He for homogenous perturbed MD (HPMD) [18] and Faussurier and Murillo for variance procedure (VP) [25], (a) for N = 256, (b) N = 1372 and at κ = 1.4.

force field strength F^* (=0.005) for the computation of the steady state values of the plasma normalized thermal conductivity is to be selected, for small varying practical system size. This feasible suitable external force field provides the steady state plasma thermal conductivity estimations, which are satisfactory over the whole range of the plasma state points (Γ, κ).

Figures 1–3 display that the computed plasma thermal conductivity is in acceptable conformity with former HNEMD investigations by Shahzad and He [13], EMD calculations of Salin and Caillol [21], inhomogenous NEMD estimations of Donkó and Hartmann [17], homogenous perturbed molecular dynamics simulations (HPMD) measurements of Shahzad and He, and theoretical predictions of Faussurier and Murillo for variance procedure (VP) [18, 25]. It can be seen from **Figure 1** that our results are slightly lower as compared to earlier known numerical results based on different numerical techniques, at lower Γ. However, the present

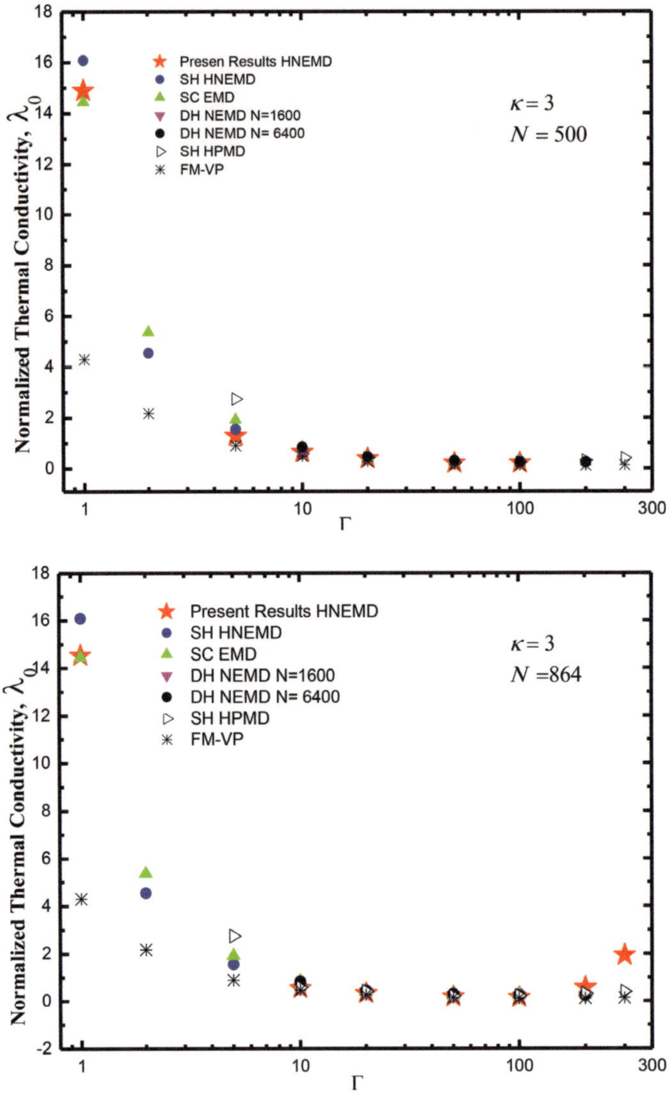

Figure 2.
Comparison of normalized plasma thermal conductivity (λ_0), computed by various numerical approaches for plasma coupling states $\Gamma \equiv (1, 300)$. Results investigated by Shahzad and He for homogenous nonequilibrium MD (HNEMD) [13], Salin and Caillol for equilibrium MD (EMD) [21], Donko and Hartmann for inhomogenous NEMD [17]. Shahzad and He for homogenous perturbed MD (HPMD) [18] and Faussurier and Murillo for variance procedure (VP) [25], (a) for N = 500, (b) N = 864 and at κ = 3.

results are well matched with earlier results for intermediate to higher Γ at two different system sizes *N* = (256 and 1372) and it is clearly shown that our results are very close EMD and HNEMD results. It is observed from **Figure 2** that HNEMD results are between EMD (at lower *N*) results and HNEMD (at higher *N*) computations at low value of Γ but our outcomes are satisfactorily matched with earlier results at intermediate and higher Γ. It can be seen from **Figure 2** that our HNEMD data for low intermediate system size it is increasing behavior at higher Γ, confirming earlier HPMD results [18]. It is observed that the deviation of data from earlier known measured data based on different methods of EMD, HPMD, and InNEMD is also computed and the outcomes of plasma λ_0 are within range of ~3–22% for EMD, ~7–20% for NEMD, and ~10–35% for HPMD. It is noted that some

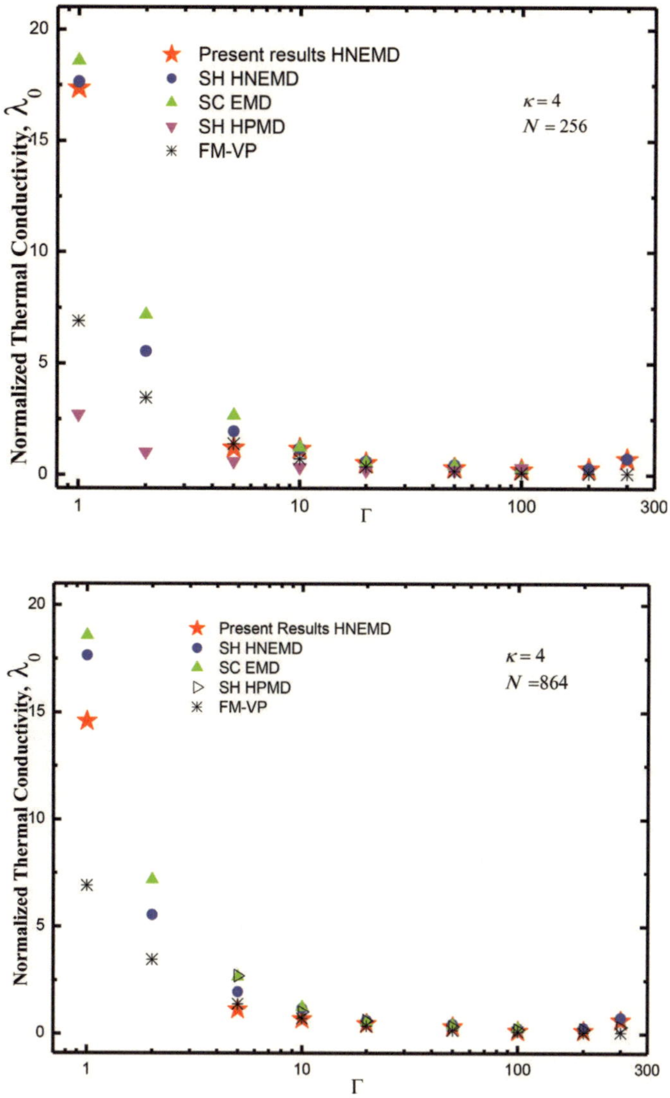

Figure 3.
Comparison of normalized plasma thermal conductivity (λ_0), computed by various numerical approaches for plasma coupling states $\Gamma \equiv (1, 300)$. Results investigated by Shahzad and He for homogenous nonequilibrium MD (HNEMD) [13], Salin and Caillol for equilibrium MD (EMD) [21], Shahzad and He for homogenous perturbed MD (HPMD) [18] and Faussurier and Murillo for variance procedure (VP) [25], (a) for N = 500, (b) N = 864 and at κ = 4.

of data points are far away from present data that are not mentioned here but most of data points are within limited statistical range, as expected. At higher screening $\kappa = 4$, it is examined from **Figure 3** that the present results are definitely lower as compared to earlier EMD computations of Salin and Caillol and HNEMD estimations at higher N of Shahzad and He. Moreover, it can be noted that the present outcomes are slightly lower at intermediate Γ and well matched at higher Γ, confirming earlier results.

It is suggested from these figures that measured outcomes are satisfactory well matched with previous outcomes at intermediate to high Γ; however, some results diverge at the lower Γ points but all within statistical unlimited uncertainty range. **Figures 1–3** show that the presented HNEMD method may precisely calculate the

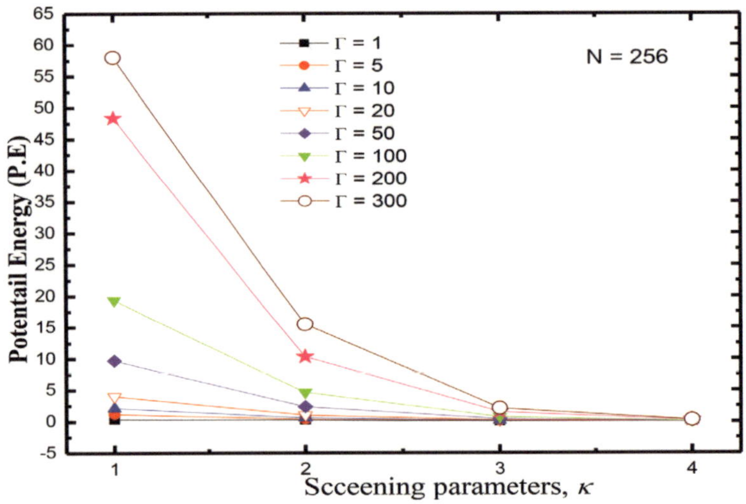

Figure 4.
Trend of normalized potential energy (P.E.) with four values of screening parameters (κ = 1, 2, 3, and 4) for different coupling states $\Gamma \equiv$ (1, 5, 10, 20, 50, 100, 200, and 300) and at N = 256.

plasma thermal conductivity of strongly coupled complex plasmas. We have shown that the present method has good performance and its accuracy is very close to earlier EMD and InHNEMD methods. It is concluded that our outcomes depend on the plasma parameters of Coulomb coupling and Debye screening strength, confirming earlier simulations. Moreover, it is shown that the position of minimum value of thermal conductivity shifts toward higher Γ with an increase of screening κ, as expected. Presently, we have demonstrated our results for a wide range of plasma parameters, ranging from nonideal gaseous state to strongly coupled range. It is noted that the extended HNMED method is excellent for lower system sizes with constant external force field strength, where signal-to-noise ratio is acceptable for equilibrium plasma thermal conductivity.

4. Summary

Plasma thermal conductivity of SCCNPs system was computed over a suitable domain of plasma couplings ($1 \leq \Gamma \leq 300$) and screening strength ($1.5 \leq \kappa \leq 4$) by employing constant external force field strength through HNEMD approach. It is shown that our HNEMD outcomes are in reasonable agreement with the earlier outcomes measured from EMD, HNEMD, InHNEMD, HPMD, and VP approaches for SCCNPs. New computations show that the minimum values of thermal conductivity exist at same values of plasma coupling Γ and it shifts toward higher Γ with an increase of screening κ, as expected in earlier numerical approaches. It has been revealed that the plasma thermal conductivity depends on plasma parameters (Γ, κ) in 3D complex dusty systems that illustrate earlier results of SCCNPs. In this study, the HNEMD method is a mostly dominant numerical approach, which occupies fast computations of plasma thermal conductivity, for small to intermediate system sizes. This chapter provides the understanding and investigation of nonlinear regime of the SCCNPs for a suitable low value of external force field strength. In future, thermal conductivity of complex plasma can be calculated by applying external magnetic field or an electric field strength and it can be applied to other systems (Coulomb, polymer, or ionic).

Acknowledgements

The authors thank Z. Donkó (Hungarian Academy of Sciences) for providing his thermal conductivity data of Yukawa Liquids for the comparisons of our simulation results, and useful discussions. We are grateful to the National Advanced Computing Center of National Center of Physics (NCP), Pakistan, for allocating computer time to test and run our MD code.

Abbreviations

SCCNPs	strongly coupled complex nonideal plasmas
Γ	Coulomb coupling
κ	Debye screening length
F^*	external force field strength
HNEMD	homogenous nonequilibrium molecular dynamics
NEMD	nonequilibrium molecular dynamics
MD	molecular dynamics
InHNEMD	inhomogenous nonequilibrium molecular dynamics
SCP	strongly coupled plasma
EMD	equilibrium molecular dynamics
λ	thermal conductivity
λ_0	normalized thermal conductivity
PBCs	periodic boundary conditions
VP	variance procedure
HPMD	homogenous perturbed MD
N	number of particles

Author details

Aamir Shahzad[1,2]* and Fang Yang[2]

1 Molecular Modeling and Simulation Laboratory, Department of Physics, Government College University Faisalabad (GCUF), Faisalabad, Pakistan

2 College of Physics, Civil Aviation University of China, Tianjin, P. R. China

*Address all correspondence to: aamirshahzad_8@hotmail.com

References

[1] Hansen JP, McDonald IR. Theory of Simple Liquids. London: Academic; 1976. Google scholar: 179

[2] Thoma MH, Kretschmer M, Rothermel H, Thomas HM, Morfill GE. The plasma crystal. American Journal of Physics. 2005;**73**(5):420-424

[3] Merlino RL. Dusty plasmas and applications in space and industry. Plasma Physics Applied. 2006;**81**:73-110

[4] Vaulina OS, Dranzhevski IE. Transport of macroparticles in dissipative two-dimensional Yukawa systems. Physica Scripta. 2006;**73**(6):577

[5] Vaulina OS, Adamovich XG, Petrov OF, Fortov VE. Evolution of the mass-transfer processes in nonideal dissipative systems II: Experiments in dusty plasma. Physical Review E. 2008; **77**(6):066404

[6] Fortov VE, Vaulina OS, Petrov OF, Vasiliev MN, Gavrikov AV, Shakova IA, et al. Experimental study of the heat transport processes in dusty plasma fluid. Physical Review E. 2007;**75**(2): 026403

[7] Shahzad A, He M-G. Structural order and disorder in strongly coupled Yukawa liquids. Physics of Plasmas. 2016;**23**(9):093708

[8] Praburam G, Goree J. Evolution of a particulate cloud in an RF plasma. IEEE Transactions on Plasma Science. 1996; **24**(1):97-98

[9] Bouchoule A, Boufendi L. Particulate formation and dusty plasma behaviour in argon-silane RF discharge. Plasma Sources Science and Technology. 1993; **2**(3):204

[10] Goertz CK. Dusty plasmas in the solar system. Reviews of Geophysics. 1989;**27**(2):271-292

[11] Rothermel H, Hagl T, Morfill GE, Thoma MH, Thomas HM. Gravity compensation in complex plasmas by application of a temperature gradient. Physical Review Letters. 2002;**89**(17): 175001

[12] Vepřek S, Reiprich S, Shizhi L. Superhard nanocrystalline composite materials: The TiN/Si_3N_4 system. Applied Physics Letters. 1995;**66**(20): 2640-2642

[13] Shahzad A, He M-G. Thermal conductivity of three-dimensional Yukawa liquids (dusty plasmas). Contributions to Plasma Physics. 2012; **52**(8):667-675

[14] Kersten H, Thieme G, Fröhlich M, Bojic D, Tung DH, Quaas M, et al. Complex (dusty) plasmas: Examples for applications and observation of magnetron-induced phenomena. Pure and Applied Chemistry. 2005;**77**(2): 415-428

[15] Feng Y, Goree J, Liu B. Accurate particle position measurement from images. The Review of Scientific Instruments. 2007;**78**(5):053704

[16] Morris D, Morriss G. Statistical Mechanics of Nonequilibrium Liquids; Australia: ANU E Press; 1990

[17] Donkó Z, Hartmann P. Thermal conductivity of strongly coupled Yukawa liquids. Physical Review E. 2004;**69**(1):016405

[18] Shahzad A, He M-G. Thermal conductivity calculation of complex (dusty) plasmas. Physics of Plasmas. 2012;**19**(8):083707

[19] Shahzad A, He M-G. Interaction contributions in thermal conductivity of three-dimensional complex liquids. AIP

Conference Proceedings. 2013;**1547**(1): 173-180

[20] Shahzad A, He M-G. Shear viscosity and diffusion motion of two-dimensional dusty plasma liquids. Physica Scripta. 2012;**86**(1):015502

[21] Salin G, Caillol JM. Equilibrium molecular dynamics simulations of the transport coefficients of the Yukawa one component plasma. Physics of Plasmas. 2003;**10**(5):1220-1230

[22] Shahzad A, He M-G, He K. Diffusion motion of two-dimensional weakly coupled complex (dusty) plasmas. Physica Scripta. 2013;**87**(3): 035501

[23] Evans DJ. Homogeneous NEMD algorithm for thermal conductivity—Application of non-canonical linear response theory. Physics Letters A. 1982;**91**(9):457-460

[24] Pierleoni C, Ciccotti G, Bernu B. Thermal conductivity of the classical one-component plasma by nonequilibrium molecular dynamics. Europhysics Letters. 1987;**4**(10):1115

[25] Faussurier G, Murillo MS. Gibbs-Bogolyubov inequality and transport properties for strongly coupled Yukawa fluids. Physical Review E. 2003;**67**(4): 046404